Phylogenetic Patterns
and the Evolutionary Process

Phylogenetic Patterns and the Evolutionary Process

Method and Theory in Comparative Biology

Niles Eldredge
Joel Cracraft

Columbia University Press
New York

Library of Congress Cataloging in Publication Data

Eldredge, Niles.
   Phylogenetic patterns and the evolutionary process.

   Includes bibliographical references and index.
   1. Phylogeny. 2. Evolution. 3. Biology—Classification. I. Cracraft, Joel, joint author. II. Title.
QH367.5.E38   575   80-375
ISBN 0-231-03802-X

Columbia University Press
New York   Guildford, Surrey

Contents

Preface

WE HAVE sought, in this book, to unite two themes too frequently disjoined in comparative biology: systematics, the ordering of the elements of the earth's biota, and evolutionary theory, the explanation of how that order arose and continues to arise. As such, it is not a "cookbook" on how to do systematics, nor does it pretend to be an overview of evolutionary theory. Rather, we have tried to characterize the basic ontological and epistemological problems of systematics and that portion of evolutionary theory relevant to systematics.

We are convinced that there is no body of theory so arcane that it cannot be discussed in simple terms. We have set out to examine the principles of comparative biology—systematics and evolutionary theory—as thoroughly and rigorously as possible. But we have endeavored to keep the language simple. This book should be accessible to both the enquiring beginning student and to the thoughtful professional. It is intended for everyone with a serious interest in the subject. We hope it is accessible to all.

We are much indebted to many people who generously helped us in our efforts to produce this book. The work seemed Sisyphean at times. Trying to write such a book at a time when comparative biology has been in an uproar has been both exhilarating and exhausting. We are grateful to Dr. Gareth Nelson, of The American Museum of Natural History, initially our collaborator, for much of the initial impetus and stimulus for the book. It was Nelson who first clarified the distinction between cladograms and trees (in an early manuscript prepared for this book). The essence of that concept is retained here, and used, no doubt, for purposes far beyond its original intent. Dr. Stanley Salthe, of Brooklyn College, has similarly provided fundamental stimulation regarding evolutionary theory, also via an unpublished manuscript dealing with hierarchical structure in evolutionary theory, as well as through numerous personal conversations. We have also benefitted from countless conversations with friends and colleagues, especially Drs. Eugene S. Gaffney, Norman I. Platnick, Bobb Schaeffer, Randall T. Schuh, and Ian Tattersall, all of The American Museum of Natural History.

We are particularly grateful to those who helped directly with production of the manuscript. Incisive, helpful commentary on various drafts and portions was received from Drs. Stephen Jay Gould (Harvard University), David Hull (University of Wisconsin, Milwaukee), Norman D. Newell (The American Museum of Natural History), David M. Raup (Field Museum of Natural History), and Stanley Salthe (Brooklyn College). Ms. Nancy A. Neff (The American Museum of Natural History and City University of New York) spotted innumerable inconsistencies and illogical formulations. Her thorough and painstaking critique of early drafts resulted in substantial improvement in many areas of the book. We also thank Mr. Bruce Manion, Ms. Sharon Simpson, and Mr. Robert Schmitz, all of the University of Illinois, for their comments on parts of the manuscript.

We are glad to acknowledge as well the patience and guidance of Messrs. John Moore and Joe Ingram, the shepherding of the manuscript through reviewers by Dr. Vicki Raeburn, and through production by Ms. Maria Caliandro, all of Columbia University Press. We are grateful to Ms. E. Penny Pounder and Ms. Majorie Shepatin for producing most of the illustrations. Finally, we thank our secretaries, Ms. Cristina Ordóñez (The American Museum of Natural History) and Ms. Claudette Lake (University of Illinois at the Medical Center), for typing some of the drafts, and also Mr. Sidney Horenstein, Scientific Assistant at The American Museum, who so ably aided us in matters bibliographic. To all, and to those unnamed, we are profoundly thankful.

New York and Chicago
January, 1980

Phylogenetic Patterns
and the Evolutionary Process

Chapter

1

Introduction—Pattern and Process
in Comparative Biology

THIS BOOK is about the twin themes of pattern and process in comparative biology. By *pattern* we mean aspects of the apparent orderliness of life. By *process* we mean the mechanisms that generate these patterns. The function of comparative biology is to analyze and capture biotic patterns and to elaborate a theory of process to explain pattern.

Comparative biology deals with several of the various sorts of patterns to be found in the biotic world. The "orderliness of life," for some, might suggest the ecological integration of individuals into populations, populations into communities, and on up through a hierarchy including provinces, realms, and finally, the entire biosphere (see Valentine 1973). A related pattern is simply the distribution of organisms in space; analyses of such patterns fall under the discipline of historical biogeography. Similarly, temporal distributions constitute the subject matter of biostratigraphy. Ecological and distributional considerations thus lead to different general approaches to the very perception of patterns in the organic world.

The kind of pattern on which we focus our attention here resides in the *intrinsic* features of individual organisms. Intrinsic features range from the atoms, molecules, and compounds composing an organism, on up through cells, tissues, organs, organ systems, as well as individual bits of behavior. Intrinsic features contrast with *extrinsic* properties of organisms—their distributions in space and time. The type of orderliness exhibited by intrinsic features is expressed in terms of relative degrees of similarity: all organisms share at least a few intrinsic properties in common (RNA, for example). Organisms tend to share a great deal more properties with certain other organisms, and share relatively little with others. Thus "orderliness in nature" in this book refers to the apparent hierarchy of similarity among the organisms comprising the biota.

To study this orderliness, we need a method, a set of procedural rules. Why is a set of rules necessary? Can we not simply map out this orderliness, which seems so obvious in a general sort of way? We take the position that the history of systematics—that branch of biology which seeks to capture the orderliness of nature—records progress from simple perception of the pattern to a conscious understanding of the nature of the order and the logical structure of the analysis of that order. The basic problem with simply "perceiving" pattern is that there is no consensus at the outset about what is to be compared, or even what "similar" means. The very fact that organisms can be viewed in a hierarchical fashion (atoms through organ systems) indicates a decided diversity in opinion about the units of comparison among organisms. And the question of what, exactly, constitutes "similarity" is perhaps the oldest one in systematics. As we shall develop, an explicit contribution to this question has been made as recently as a quarter-century ago (Hennig 1950). In our view, this contribution is fundamental and effects any formulation of a logical structure for pattern analysis in comparative biology.

Thus a fundamental cornerstone of comparative biology is the simple assertion that there is order in the biological realm of nature—an order involving patterns of similarity among the constituents of the earth's biota. Those who do not perceive such order, or who refuse to adopt its existence as an axiom, will, of course, have little interest in the subject matter of this book.

How did this pattern of similarity of features originate? Are there general mechanisms which operate in the natural world that are capable of generating this pattern? There are two general sets of explanations for this pattern still with us in the twentieth century: evolution and separate, or special, creation. Evolution asserts that the pattern of similarity by which all known organisms may be linked is the natural outcome of some process of genealogy. In other words, all organisms are related. Just as human children resemble their parents, organisms which are closely related tend to resemble each other more closely than do their more remote relatives. The hypothesis that such a process of genealogical production of descendants from ancestors has occurred is, of course, called "evolution." It is the only generally accepted biological mechanism for the production of this pattern, ever since the demise of the notion of spontaneous generation long ago.

Special creation, in its various guises, is not a scientific formulation, simply because it asserts that every discrete "type" of organism has its own unique beginning unconnected with the history of any other kind of organism. All organisms are created *de novo*. Under such an assumption, there are no generalizations to be found which would cover more than a single case. The most common form of special creation today is associated with the religious concept of a supernatural power, or being, which has created the order we see in the biotic world. Such formulations have always been difficult to investigate experientially. Special creation is, properly speaking, a set of assumptions about the nature of the world that excludes such experiential considerations. Yet, if a notion is to be considered within the purview of science, it must be capable of evaluation through our experiences with the world. Those who adopt the assumptions and axioms of special creation will not be inclined to adopt the views presented in this book.

The notion that life has evolved paves the way for a general theory of mechanisms to explain the genesis of life's orderliness. While the concept that life has evolved remains the only viable alternative to special creation and allows the scientific search for general mechanisms, evolutionary theory per se is actually in a healthy state of disrepair. Unlike systematics, where methodologies for reconstructing the pattern of nature form the very basis of theoretical discussion, evolutionary theorists have concentrated almost totally on generating and examining statements about how the process works, and have spent almost no effort in evaluating just how the process of evolution ought to be studied scientifically. We address both epistemological and ontological questions in evolutionary theory in this book.

Adopting the assumption that life has evolved as our other cornerstone of comparative biology allows us to expand the concept of pattern, or "order in nature." Putting the two together, it is clear that the pattern is a direct result of the process, i.e., that the pattern is historical. Whether the pattern is being "perceived," or "analyzed," it is the contribution to our understanding of the history of life that is actually at stake.

Pattern and Process: Logical Interconnections

Pattern results from process. Beyond this simple assertion, we might ask what, if any, other relationship might exist between these two areas of systematic biology. The question has many possible answers, all lying in the methodological, or epistemological, realm (see Bretsky 1979; Eldredge 1979a; Eldredge and Cracraft 1979). It is apparent that most systematists do not feel the need to adopt an explicit set of notions about evolutionary processes in order to pursue the task of reconstructing the history of life. Reference to the 27 volumes (at this writing) of *Systematic Zoology* amply demonstrates how little a concern for evolutionary mechanisms characterizes normal work in systematics. Likewise, reference to such journals as *Evolution* or *American Naturalist* reveals of how little concern evolutionary historical patterns—or, *phylogenetic* patterns—are to those who are primarily concerned with elaborating ideas of evolutionary mechanisms. Yet, delving a bit more deeply, it becomes clear that virtually no one operates in either area without at least some reference to the other. In sum, all systematists working on the analysis of pattern have some set of assumptions at least subconsciously concerning the evolutionary process, and these sets are inferable from the nature of the "pattern" they perceive and report in the literature. Conversely, except perhaps for the purely inductive mathematical treatments of some population geneticists, all evolutionary theorists have a general concept of the pattern they are trying to explain and, on a more basic level, usually have some sort of historical data on which to base their analyses (see chapter 6). Thus, the connections between the two areas of comparative biology are deep, if not always clearly acknowledged.

It is our position that, in the analysis of evolutionary history (i.e., pattern) in its most general form, we need only adopt the basic notion that life has evolved. Only when more detailed statements are required are more specific notions of evolutionary processes relevant. We believe that the most important connection between the two areas, an aspect as yet underexplored, involves the comparison of the patterns of both intrinsic and extrinsic features of organisms predicted from theories of process, with those actually "found" in nature. Initially, therefore, the study of pattern must be divorced as much as possible from the study of process, to provide an unbiased baseline for the evaluation of alternative hypotheses about process. In discussing methodological questions in both areas of comparative biology, we have adopted the view that the procedures should be hypothetico-deductive in nature: elaborate a hypothesis which contains the basis for its own evaluation (i.e., predictions). In this approach, nothing is ever "proven;" we speak, rather, in terms of elimination ("refutation" or "rejection") of manifestly false hypotheses, and retention of as yet unfalsified hypotheses. Hypotheses consistently resisting rejection relative to alternative hypotheses are said to be more highly corroborated. Under this view, facts themselves are nothing more than highly corroborated hypotheses.

The application of the hypothetico-deductive approach to systematics has been much discussed of late (see Gaffney 1979, for a thorough review and guide to the literature). In keeping with the general disinclination of those interested in evolutionary mechanisms to be concerned with methodological questions, the subject has been barely discussed in this branch of comparative biology. As we shall discuss at various stages throughout the ensuing chapters, the rabbit warren of untestable story-telling ("scenarios") which comprises much of past and contemporary evolutionary theory indicates that future work in this area could benefit greatly if it were cast more explicitly in hypothetico-deductive terms. It is our position that one way we might accomplish the task of making evolutionary theory more explicitly hypothetico-deductive is to use highly corroborated hypotheses of phylogenetic pattern to evaluate predictions of pattern generated from hypotheses about the nature of the evolutionary process. Thus, in our opinion, improvement in evolutionary theory will depend, to a great extent, on the availability of highly corroborated hypotheses of evolutionary history. And such hypotheses of pattern depend upon the adoption of rigorous methods of phylogenetic analysis.

The Study of Phylogenetic Pattern

The biological discipline known as systematics deals with the theory and practice of capturing the orderliness in nature that has resulted from patterns of phylogenetic ancestry and descent. There are two closely related general concepts pertaining to phylogenetic patterns. First, a notion of evolution implies "descent with modification"; new intrinsic features, be they genes, specifiable anatomical structures, or bits of behavior, arise from time to time and are inherited by descendants. This is the general idea of the origin and maintenance of morphological diversity among organisms. Because some new features appear earlier than others in evolution (the amniote egg before mammalian hair, for example), the expected outcome is a nested set of evolutionary resemblances: *similarities among organisms are hierarchically ordered as the expected outcome of the evolutionary process itself.*

The second concept, a corollary of the first, involves the hierarchical arrangement of *taxa*. Taxa are groups, or sets, of organisms, defined and recognized according to a set of criteria. The primary task of systematics is the recognition and classification (naming sets within a hierarchical arrangement) of taxa. The two concepts of nested similarities and nested taxa are closely related, both epistemologically (i.e., methodologically) in systematics, and both ontologically and epistemologically in evolutionary theory. All approaches to understanding the hierarchical arrangement of taxa in systematics depend on the nested pattern of similarity of intrinsic features as the primary data for analysis: the nested sets of features reveal the outline of the nested set of taxa. As far as evolutionary mechanisms are concerned, the relationship between nested sets of intrinsic features and nested sets of taxa is a complex issue, having to do with such fundamental problems as the very definition of biological evolution itself. For the moment, we shall summarize the spectrum of opinion regarding this relationship by examining the conventionally recognized "schools" of systematics.

It is generally said that there are three contemporary "schools" of systematics: evolutionary systematics, phylogenetic systematics (also known as cladism or cladistics), and numerical taxonomy (generally, if inaccurately, used as a synonym of "phenetics"). None of these three schools is monolithic; in terms of the results of their analyses of identical data sets, it might even be claimed that they are not very much different at all. But there are general theoretical stances attributable to each which summarize the spectrum of philosophical positions taken by contemporary systematists.

Both numerical taxonomy, the most vigorous modern manifestation of pure "pheneticism," and phylogenetic systematics can be viewed as late-coming developments that at least partly represent reactions against evolutionary systematics. Cast in the role of the "traditional approach" (largely by virtue of the vocal assaults, first by numerical taxonomists, then by cladists, and the equally assertive reactions by its generally acknowledged spokesmen), evolutionary systematics is actually a multifarious discipline which last received a thorough intellectual housecleaning in the 1940s. The "new systematics" is characterized particularly by an attempt to take intraspecific variation into account, with a concomitant eschewal of "typology" (i.e., the characterization of taxa, especially, but not exclusively, species, as if they did not vary). It was widely hailed as a great advance, in keeping with the spirit and letter of the synthesis in evolutionary theory, itself wrought in part by the very same biologists. It would appear more accurate to view evolutionary systematics as another, equally vigorous area of systematics, rather than simply and cavalierly as a body of thought whose time has come and gone simply because of its imagined advanced age.

Although all three "schools" of systematics are difficult to characterize briefly without caricature, evolutionary systematics is especially difficult. It is tempting to suggest that this state of affairs directly reflects the fact that the much-vaunted "synthetic theory" of evolution, so closely tied to this approach to systematics, is a great deal less completely "synthesized" than is popularly imagined (we expand on this theme later on in this chapter and at length in chapter 6). In any case, there does seem to be a core of basic propositions probably agreeable at least to the majority of the practitioners of evolutionary systematics.

The central tenets of the evolutionary school of systematics seem to be that life has evolved, that the order we see is a product of evolution, and that the goal of systematics is to reconstruct that evolutionary history as closely as possible. Furthermore, evolution implies ancestry and descent, and therefore the hierarchical nature of this order should be depicted on phylogenetic trees, which specify patterns of ancestry and descent. Descendants genealogically far removed from ancestors should resemble each other less than more closely related ancestors and descendants, reflecting, at base, degrees of genetic similarity. Thus in the evolutionary school there are two different *kinds* of similarity: (a) true evolutionary resemblance, i.e., resemblance due to inheritance from a common ancestor, and (b) false resemblance (usually termed convergence; "paralellism" is included here, but is regarded by most evolutionary systematists as a sort of intermediate case between "true" and "false" resemblance). False resemblance does not reflect inheritance of the similar structures from a common ancestor. Trees derived from such analyses can then be converted into a classification by adopting a set of correspondence rules (to use the phrase of Colless 1977).

Not all evolutionary systematists necessarily accept all these propositions. Moreover, there is a great deal more to formulations of this school than the crude, bare outline above, as readers of Simpson's (1945, 1961) and Mayr's (1969) books on the subject will attest. But the characterization we have given does at least seem to summarize the basic propositions in a fashion probably more or less agreeable to all.

The late 1950s and the decade of the 1960s saw the development of numerical taxonomy. With the publications of *Principles of Numerical Taxonomy* (Sokal and Sneath 1963; a second book, updating the first, appeared as Sneath and Sokal 1973), the theoretical basis of numerical taxonomy was established. The main concern appeared to be a desire to reformulate the process of delineating life's orderliness in a more standardized, repeatable, rigorous, and objective fashion (Sokal and Sneath 1963:49). In a sense, this approach treats the pattern as if it merely needs to be "perceived," as if it were like any collection of "data." Viewed in this manner, the scientific perception of the pattern is to be formalized and made objective like any other precise, but routine data gathering operation in science. This view is far removed from the general notion that any "observation" of nature constitutes an hypothesis.

Numerical taxonomists note the many difficulties in reconstructing phylogenetic history, and therefore tend (e.g., Sokal and Sneath 1963) to dissociate their activities from phylogeny reconstruction. Acknowledging that the patterns do reflect phylogenetic history, numerical taxonomists nonetheless have claimed that such history is unknowable with certainty, or in fact in any detail. Objecting particularly to Simpson's (1961:110) remark that "like many other sciences, taxonomy is really a combination of a science, most strictly speaking, and of an art" (a statement which pertains to the construction of classifications only, and not to phylogeny reconstruction), numerical taxonomists sought to make systematics more "scientific."

The central methodological principle in numerical taxonomy is *phenetics*—the clustering of samples ("operational taxonomic units," or "OTUs") according to an index of overall similarity. "Phenograms," or "dendrograms," are generated (almost always by computer) according to an algorithm specifying for one of the many available measures of similarity among OTUs. Tracing their ideological pedigree back to the botanist Adanson (1763), who advocated examination of as many characters as possible and the production of classifications based on overall similarity, numerical taxonomists contrasted their "objective" approach with the "subjective" weighting (selectivity) of characters practiced, they claimed, in a rather vague and arbitrary fashion by evolutionary systematists.

Thus, concepts of similarity and its measurement seem to constitute an important difference between pheneticists and evolutionary systematists. The concept of overall similarity consciously lumps "true" evolutionary resemblance with convergent and parallel resemblance: the position of most numerical taxonomists seems to be that, if enough characters are examined, "real" resemblance will outweigh "false" resemblance. Evolutionary systematists, on the other hand, make explicit the distinction between "true" (i.e., evolutionary or phylogenetic) and "false" (convergent) resemblance. However, once the "true" set of resemblances is identified, the operation is purely phenetic: the basic criterion of evolutionary systematics, at least according to Mayr (1969:200) and Bock (1977) is the maximization of genetic similarity as judged by the phenotype. We conclude that the difference between numerical taxonomy and evolutionary systematics, great as the methodological differences seem to be, boils down to slightly different views about what is attainable insofar as the scientific analysis of life's orderliness is concerned. Evolutionary systematists strive to make trees, with ancestors and descendants, while numerical taxonomists are content with statements of relative "nearness" of samples, and thus define taxa directly on the basis of clusters found on phenograms generated by some measure of overall similarity.

The third school, "cladistics," or "phylogenetic systematics," seems, in many respects, to occupy an intermediate position on these various issues. With the evolutionary systematists, cladists believe that the orderliness of the biotic world derives from evolution, and that the reconstruction of the history of life is the central goal of systematics. It was Hennig (1950, 1966) who first made explicit yet a third component of "similarity." He accepted the usual dichotomy between "true" evolutionary and "false" similarity, but further pointed out that, for any monophyletic taxon (i.e., a taxon composed of two or more species consisting of an ancestral species and all its known descendants), evolutionary similarities shared by a group are of two sorts: those held over from some remote common ancestor (e.g., two pairs of limbs in mammals) vs. those held only by members of that group (e.g., three inner ear bones in mammals). Hennig pointed out that older evolutionary novelties can be retained in a sporadic manner; consequently their utility for defining and recognizing clusters of organisms is appropriate only to the hierarchical level at which they represent true evolutionary novelties. Thus cladists amplify the set of kinds of similarities recognized by evolutionary systematists, seemingly departing even further from the position of the pheneticists.

However, because ancestors never possess a set of evolutionary novelties unique to themselves, their definition and recognition is, logically, difficult. For this reason, cladists concentrate on a more elemental level: the prime goal of systematics, according to cladists, is the definition and recognition of monophyletic groups. This is accomplished by the search for nested sets of evolutionary novelties depicted on branching diagrams called "cladograms." Cladograms order organisms according to nested sets of these novelties; consequently, the organisms are ordered as well into nested sets— (hypothesized) monophyletic taxa (see chapter 2). The procedure has the added advantage of being easily converted into classifications with a minimum of required conventions (see chapter 5).

As diagrams of the history of taxa, cladograms can be interpreted in terms of relative recency of common ancestry. As first pointed out by Nelson (personal communication, 1976), and further characterized by Platnick (1977b), Tattersall and Eldredge (1977), Cracraft (1979), and Eldredge, (1979a) (see Wiley 1979 for a counter opinion), a cladogram subsumes the logical structure of a set of trees. Phylogenetic trees, in specifying actual series of ancestral and descendant taxa, are more detailed and precise sorts of hypotheses than are cladograms. Thus, as Platnick (1977b:441) has discussed, cladists adopt a less extreme view than evolutionary systematists, in that their scientific goals regarding the reconstruction of phylogeny are tempered by the theoretical and methodological difficulty of dealing with ancestors. On the other hand, the cladists do not share the utter despair that numerical taxonomists have expressed concerning the impossibility of dealing with the history of life in a rational manner. We should adopt the position (see also Eldredge 1979a; Wiley 1979) that, logically, the ancestral units of evolution are species—i.e., that supraspecific taxa do not form ancestral-descendant units. Therefore any system, be it a branching diagram or a classification, which recognizes supraspecific taxa as ancestors is illogical. In this sense, there is no difference between a cladogram and a tree depicting relationships among supraspecific taxa, inasmuch as the added information of trees (identification of certain taxa as ancestors) is superfluous. However, species do form ancestral-descendant sequences, and thus, logically, their histories are appropriately depicted on phylogenetic trees. We discuss the methods by which a cladogram can be converted into a phylogenetic tree for species in chapter 4 and further claim, in chapter 6, that phylogenetic trees are the actual patterns required for the scientific study of speciation.

Again, there seems to be little difference between cladists and evolutionary systematists (and numerical taxonomists, for that matter) except for a different emphasis on such issues as the nature of scientific inquiry in systematics and what might accordingly be attained in terms of the reconstruction of the history of life. Both Mayr (e.g., 1974:98) and Simpson (1975:14), as well as other noted evolutionary systematists, have acknowledged the validity and importance of Hennig's explicit statement that the evolutionary process produces, as an expectation, a nested set of evolutionary novelties. The additional information that evolutionary systematists wish to incorporate into their trees and classifications seems to be based on certain assumptions about the evolutionary process, particularly the extreme importance accorded to adaptation as the central theme and problem in evolution (see chapters 5 and 6 for an extended discussion of this issue).

But, at the elemental level of recognition of patterns of similarity, we might fairly ask if there are any fundamental differences in addition to the obvious methodological ones among the three schools. Pheneticists talk of overall similarity, evolutionary systematists speak of overall (genetic) similarity in the context of "true" (as opposed to "false") evolutionary similarity, while cladists additionally recognize levels of "true" similarity. Cladists avoid the confusing issue of "weighting" by recognizing that all nonconvergent characters are relevant to defining monophyletic groups at some level. The problem is the recognition of the correct level for any character (see chapter 2). But congruence of patterns of similarity can be expected, in many cases, to lead to identical groups, whether the analysis is performed by a numerical taxonomist, a cladist, or an evolutionary systematist (Nelson 1979).

There is, however, one fundamental difference between these three approaches to pattern analysis in systematics. The difference pertains to the definition of taxa, and not to the analysis of similarity per se. Without an explicit attempt to evaluate similarities at their proper hierarchical level, clusters of organisms whose similarities are primitive retentions rather than evolutionary novelties typically result. Thus the problem with phenetics in general is not parallelism or convergence (a problem usually resolvable, especially with reference to a parsimony criterion—see chapter 2) but, as in evolutionary systematics, the failure to evaluate evolutionary novelties at their proper level. The effect this has had on analyses in both evolutionary systematics and numerical taxonomy is fundamental: some taxa so defined are non-monophyletic. Both numerical taxonomists and evolutionary systematists regularly admit groups based on shared retention of primitive features into their systems. Such groups are inevitably of dubious cohesion, as the evolutionary systematists themselves seem to admit when they endorse the use of evolutionary novelties for the definition of taxa. As Farris (1977) and Platnick (1978b) have recently discussed, monophyletic taxa maximize comparative information about organisms. Moreover, and this argument is crucial to all considerations of evolutionary process in this book, *evolution is first and foremost a genealogical process. It produces genealogies of ancestors and descendants. If we are to compare pattern with theories of process—as we must do to improve our very notions of process—we must have at the very least an accurate concept of evolutionary genealogy.* As will be developed in chapter 6, it is a mistake to invent theories of process to explain the origin of non-monophyletic groups. It follows, then, that any procedure in systematics whereby nonmonophyletic groups are routinely recognized distorts not only the information content of the classificatory system, but also our very notions of evolutionary processes. This suggests that the system best suited to the recognition of monophyletic taxa is the best system for comparative biology as a whole.

The Study of the Evolutionary Process

As we stated earlier in this chapter, most evolutionary theory is onto-logical, rather than epistemological; there is far more concern with how the process works in nature, than with how we know about that process. Many of the problems in evolutionary theory stem from an inattentiveness to the "how we know" component. Specifically, there is a very strong tendency, amounting to a tradition, to (a) reconstruct phylogenetic trees, (b) elaborate some theory of evolutionary mecha-nisms, and (c) subsequently explain the phylogenetic patterns of (a) in terms of the theory of (b). (See, for example, the statement to this effect by Bock and Von Wahlert 1963:140.) Not only do theories tend to be invented which can explain all patterns (a frequent and often justified complaint about the use of the concept of adaptation by nat-ural selection, for example), but, perhaps more insidiously, the pat-terns themselves are perceived in such a fashion as to fit prevailing notions of process. As an example, consider the popularity of recog-nizing polyphyletic groups in the late 1950s and early 1960s. Ap-parently stemming from Huxley's (1958) paper discussing grades and clades (patterns) as well as anagenesis, cladogenesis, and sta-sigenesis (modes of process producing the patterns), "polyphyly" enjoyed wide discussion as an evolutionary process. In our view, this flurry of theoretical work represents a sort of hypertrophication of the general view that evolution is fundamentally the transformation of intrinsic features, best explained by reference to adaptation *via* natu-ral selection. Investigators cheerfully recognized the existence of taxa acknowledged *not* to share unity of descent—e.g., Mammalia, once claimed by Simpson (1959), to have arisen nine different times, a view no longer in favor. Rather these taxa were based on con-vergences, parallelisms, and, to a lesser extent, joint retention of primitive features. (The theme definitely emphasized independent acquisition of evolutionary novelties designed to perform the same function.) This celebration of adaptation came directly from evolutionary theory and, inasmuch as it dwelt on the "explanation" of the evolution of phylogenetically non-existent groups, is best regarded as a bizarre conclusion resulting from the application of a theory far beyond its appropriate limits (see chapter 6).

It is our firm conviction, in contrast, that the conventional procedure of explaining pattern in terms of notions of process (step 3 above) is grossly in error. We agree that phylogenetic patterns must be analyzed (as wholly independently from notions of process as humanly possible) and theory invented to explain pattern—i.e. the first two steps. But the crucial third step should be the direct comparison of predictions of pattern drawn from ideas of process directly with the analyzed phylogenetic patterns themselves, with the aim of critical evaluation of the notions of process themselves. A major problem of contemporary evolutionary theory from an epistemological standpoint is not so much that its propositions are untestable, but that its main practitioners use theory to explain away pattern (see Grene 1959, for a clearly analyzed example). Rather, the theory should be tested by comparing it with best estimates of actual evolutionary *results*—i.e., carefully reconstructed patterns of phylogenetic relationship.

From an ontological standpoint, there is also much to criticize in contemporary evolutionary theory. As systematists, our main concern is with patterns of relationship among taxa. Systematists naturally have been most conversant with, and have themselves helped to create, that part of evolutionary theory specifically addressed to problems of the origin, persistence, and extinction of taxa. Most recently, this work has focused on species (e.g., Mayr 1942, 1963), and for good reason: species are unique as taxonomic entities (see chapter 3). In discussing among-taxa evolutionary problems, a very subtle but crucial issue is immediately raised. Just as cladograms depict nested sets of characters, and thereby also depict nested sets of monophyletic taxa, so does the issue of the evolution of taxa become intertwined and confused with the modification of intrinsic features in evolutionary history. We can define taxa only in terms of these intrinsic features. It was probably inevitable that the predominant view of evolution would stress the *transformation* of intrinsic properties as the central issue. There is no doubt that this has happened (chapter 6). That transformation of intrinsic properties is the central theme of evolution to most biologists is evident in the general definition of evolution historically acceptable even to prominent systematists (e.g., "evolution is any change of gene content and frequency within populations"). Because we characterize taxa by intrinsic properties, we tend to assume that their evolution is a matter solely of the transformation of these features rather than a matter of the origin of species as well. This reductionist viewpoint has been paramount and underlies the lack of true synthesis in evolutionary theory.

In contrast, we have noted (especially in chapters 3, 4, and 6) a historical thread of interest in problems of the evolution of taxa which is to be sure, related to, and intermixed with, *but not the same as,* the problem of the transformation of intrinsic features. Taxonomic diversity, in our opinion, is not a synonym of morphologic diversity. They are related, and *how* they are related is an interesting problem. But they are not the same. The bulk of contemporary evolutionary theory focuses on morphological diversity, to the point, at times, of presuming that the two sorts of diversity are isomorphic (see chapter 6 for an extended discussion of these issues). In keeping with our initial perspective as systematists, we have adopted and defend throughout this book the position that species are real entities existing in nature, whose origin, persistence, and extinction require explanation. All species (except those reproducing strictly asexually, with no exchange whatsoever of genetic materials among individuals—a distinct minority of organisms) are held together by a pattern of parental ancestry and descent that is disrupted when new species arise from old. Adaptation and natural selection are hypotheses to explain changes of intrinsic features within populations from one generation to the next. Speciation disrupts these lineages of parental ancestry and descent. *The foregoing implies that among-species differences do not flow directly as a simple, reductionist extrapolation of within-population generational change. The central role of species as real units in nature—these units being the ancestors and descendants of the phylogenetic process—implies a view of distinct phenomenological levels in the evolutionary process.* In such a view, microevolution is, perhaps, best defined as change in gene content and frequency within populations, and macroevolution is defined as change in species composition within a monophyletic group in space and time, best thought of, perhaps, as a process of differential species origination and survival within monophyletic taxa. We explore these possibilities in far greater detail in chapters 3, 4, and especially 6.

Integration of Pattern with Process: The Structure of This Book

The structure of the remainder of this book can be summarized as follows: First, in chapter 2, we discuss method and theory pertaining to the analysis of evolutionary novelties: cladistic analysis. It is this nested set of intrinsic features which gives us the main signal of interrelationships among taxa. It is through cladistic analysis that we derive our hypotheses of (a) the composition of monophyletic taxa and (b) the distributions of nested sets of characters in the biotic world. Both types of sets have further uses.

Next, in chapter 3, we discuss the special case of species. We define species in such a way as to stress their internal cohesion, their identity as discrete, real entities, and their unique position as phylogenetic units. No taxon other than species serves as ancestors and descendants (i.e., as phylogenetic units) in evolution. We discuss two sets of criteria that form the basis for the evaluation of hypotheses of species composition.

Then, in chapter 4, we briefly discuss basic hypothesized modes of speciation (both those consistent with our own characterization of the nature of species, as well as some consistent with other species definitions). This departs from our overall plan to use pattern to evaluate predictions derived from theories of process. We follow this procedure to expose the theoretical patterns of possible phylogenetic trees. We then discuss trees themselves, finding them logical structures only when species are considered, and adduce some methodological rules for converting cladograms into trees.

Chapters 5 and 6 constitute our effort to characterize how branching diagrams of interrelationships among taxa may be put to further use. In chapter 5, we detail the methods and procedures, as well as history of ideas and general purpose, of the construction of biological classifications. We conclude that the history of systematics might best be characterized as the progressive elimination of false ("unnatural") sets, especially those sets "defined" by the absence of features used to define a coordinate set; we refer to these sets as "not-A" (e.g., Invertebrata) and "A" (e.g., Vertebrata) sets, respectively. We conclude that monophyletic groups maximize the information content of a classification and, because ancestral taxa cannot be monophyletic (by any acceptable definition of monophyly), they therefore cannot be arranged in an hierarchic fashion without special conventions. We conclude, therefore, that cladograms alone are necessary and sufficient for the construction of classifications. We present some correspondence rules for the conversion of cladograms into classifications.

Finally, in chapter 6, we consider the use to which highly corroborated hypotheses of patterns of relationship among taxa might be put in the evaluation of statements in evolutionary theory. We take the position that the majority of evolutionists (whether neo-Darwinists, syntheticists, or saltationists), have focused on the explanation of change of intrinsic features, primarily on morphologic change. The most common mode of explanation is couched in terms of the paradigm of adaptation via natural selection. We point out that the extrapolation of within-population dynamics of variation and selection to explain morphological and adaptational differences among taxa of higher categorical rank results in a theory of the evolutionary process which purports to discuss taxa but in reality only deals with the intrinsic properties of organisms. We conclude that if species are discrete reproductive units, microevolutionary processes cannot logically be extrapolated, in a reductionist manner, to explain macroevolutionary patterns.

We then examine evolutionary theory from the standpoint of phenomenological levels. We conclude that microevolution (within-population, within-species transformation of intrinsic features) may be hypothesized to result from adaptation via selection. For its scientific study, microevolution requires direct historical information, on a generational basis, which can be obtained experimentally. Speciation itself requires detailed phylogenetic trees for its scientific study. For the analysis of macroevolutionary patterns—i.e., patterns of origin of, and fluctuations of diversity within, monophyletic taxa of rank higher than species—only cladograms are required, which supply

(a) the nature of the composition of the monophyletic groups, and (b) relative distributions of evolutionary novelties. Each component (i.e., microevolution, speciation, macroevolution) of evolutionary theory relies upon use of the kinds of historical data (hypotheses) appropriate to it, for its eventual improvement. Such improvement is to be expected particularly when predictions of pattern, based on theories of process, are directly compared with independently derived hypotheses of pattern, in a hypothetico-deductive manner. We close the chapter, and the book, with a brief characterization of a testable, taxically oriented theory of macroevolution.

Chapter
2

Cladograms: Cladistic Hypotheses and Their Analysis

THE SUBJECT of this chapter is a fundamental one in comparative evolutionary biology: how do we reconstruct the history of life? We will claim that hypotheses about phylogenetic history are necessary standing points for virtually all comparative evolutionary studies. Unfortunately, many evolutionary biologists either choose to ignore the problem of phylogenetic history altogether, or it is dismissed on the grounds that "direct" evidence is lacking—the implication being that there is no fossil record for the group in question. In this chapter we will suggest that an estimation of phylogenetic history can be obtained with or without a fossil record (although paleontological data are surely welcome in any study), and subsequent chapters will elucidate the importance these estimations have for investigating evolutionary questions. To many biologists the claim that phylogenetic history is capable of analysis without paleontological evidence may sound a dissonant chord, but the problem can be resolved by understanding not only the general philosophical approach adopted in this book but also the specific kinds of phylogenetic hypotheses we believe are subject to scientific investigation.

In our view, any methodology designed to reconstruct phylogenetic history should be hypothetico-deductive in structure, that is, alternative hypotheses must imply different empirical consequences and these in turn serve as the basis for evaluating those hypotheses. Thus, each hypothesis is expressed so that it can be criticized and rejected if the evidence so warrants. We are interested, therefore, in a methodology capable of producing and evaluating hypotheses of this kind. We reject outright any predilection for narrative-type scenarios of phylogenetic history if they are not expressed in a rigorously testable form. The literature is replete with these narratives, many of which may sound very reasonable, but all too frequently there is little or no way to subject them to critical analysis. In such cases, the descriptions of historical events essentially lie outside the realm of scientific inquiry.

In this book, we are concerned primarily with erecting hypotheses about the *pattern* of life's history, including the supposed sequence of phylogenetic branching events and the distribution of similarities and differences among the organisms being studied. Thus, the initial, primary question is exactly *what* has happened, not why or how. These hypotheses, then, should attempt to explain, as simply as possible, all the relevant empirical data bearing on the question of what has been the history of life.

Cladograms, as defined and discussed here, are specific kinds of hypotheses about the history of life. They are hypotheses about pattern. The concept of evolution implies change or modification in the intrinsic properties of organisms from some state or condition, *a,* to a derivative state or condition, *a'*. More significantly, however, evolution means that somehow one kind of organism may change or evolve into a different kind of organism. In other words, there is ancestry and descent. It follows that similarities in intrinsic properties among organisms are, in some way, a manifestation of the process of evolution. From Darwin to the present the evolution of life has been viewed as a process of branching and diversification. Such a process produces a pattern, and that pattern is hierarchical, or, put another way, it is a sequence of nested sets of organisms (the "groups within groups" of Darwin), and it is generally agreed that the similarities among these organisms must mirror that hierarchy in some manner. Hence, systematists look to the analysis of similarities to give some understanding of that hierarchy.

The question arises as to the kind of similarity that might be expected to evidence this nested pattern. After all, organisms may be similar to one another in various ways. Does this nested pattern reflect a general measure of the overall resemblance of one organism to another? Or is this similarity of a more special, more specific kind? The answer is related to the process—evolution—that has produced the pattern of branching and divergence. The assumption of evolution implies that while there is a continuity in the characteristics of ancestor and descendant, there is also change. During the evolutionary process, newly evolved organisms come to be characterized by novelties, their descendants retain these, and thus there is an expectation of nested sets of evolutionary novelties.

Cladograms, therefore, are hypotheses about the pattern of nested evolutionary novelties postulated to occur among a group of organisms. As such, cladograms are branching diagrams of organisms or groups of organisms. Somewhat more formally, one can view the hierarchical structure of cladograms in terms of groups (sets) and subgroups (subsets): the elements (groups of organisms) sharing a branch point comprise a subset within other subsets, themselves defined by higher branch points of the hierarchy. Hence, cladograms are representative of our knowledge about organisms in that as the hierarchy is ascended, that is, as more and more branch points are included, statements about this knowledge can be said to become more general. Cladograms, therefore, denote levels of generalization, and thus specify different hierarchical levels. Statements about attributes pertaining strictly to cats, for instance, are less general (more specific) than those pertaining to all mammals, and the latter, in turn, are more restricted than those pertaining to all vertebrates.

The question arises: to what extent are cladograms to be considered actual representations of phylogeny? The concept of the cladogram has been synonymous with "phylogeny" in some of the prior literature. The view taken here is that cladograms, in themselves, are not phylogenies, but rather hypotheses about the pattern of nested evolutionary novelties. This pattern allows us to arrange organisms into sets and subsets. Phylogenetic trees specify ancestry and descent, whereas cladograms do not. Thus, the construction of phylogenetic trees requires certain evolutionary assumptions which themselves are not necessary for cladogram construction. The distinction between cladograms and trees will be examined further in chapter 4.

Cladograms: Some Introductory Principles

A Simple Example

In order for the reader to gain some initial insight into the notion that cladograms express our general knowledge about organisms and are constructed in terms of nested evolutionary novelties, it may be helpful to turn first to an example using organisms familiar to most, if not all, biologists. This example is presented in part to demonstrate that similarities and differences among organisms can be expressed by a nested hierarchy, and the example itself will be used to formulate a series of general statements about cladograms. Naturally, because the example is a simple one, it cannot convey the complexity of problems often confronting a systematist undertaking a cladistic analysis. Nevertheless, the exercise should help to formalize the methods of many systematists, methods that until recently have often remained essentially intuitive.

The example is taken from the vertebrates. Only five well-known kinds of organisms are considered, and a small, restricted sample of the characters possible has been chosen. (At this point in the discussion little would be gained by adding organisms or characters: how the investigator might handle more complicated situations will be outlined later in the chapter.)

The problem at hand is to construct a cladogram for the following organisms: cat, mouse, lizard, perch, and lamprey. Suppose each organism is represented by specimens encompassing all phases of its life cycle. Because these are preserved specimens, observations that might have been made on living material, for instance, physiology, biochemistry, or behavior, are not a matter of immediate concern. For the moment, the analysis can be restricted to the morphological features in the available specimens.

The first task is to undertake a survey of the anatomy of these organisms and to note the similarities and differences. This morphological comparison indicates that the five organisms exhibit a broad structural diversity. Moreover, each organism shows considerable variability in form from one stage of its life cycle to another. Because a number of specimens of each life history stage are available, variability in some features is also observed among specimens of the same stage. Taking all this into account, and assuming the distinctness of each organism, how are the observed similarities among the five organisms to be evaluated so that a cladogram may be constructed?

To answer this last question, we might consider one of two approaches. First, the analysis might be confined to the five organisms; in this case an attempt must be made to form nested groups by nesting statements about evolutionary novelties postulated from the similarities observed among the five organisms. In the second approach, a comparison with other groups of organisms might be used to suggest which similarities appear to be evolutionary novelties and thus what the pattern of groups and subgroups might be. As will be developed later in the chapter, there is reason to believe both approaches will lead to similar results, namely, a cladogram that depicts our knowledge of the organisms in terms of nested sets of evolutionary novelties.

Upon dissection and study of the specimens a number of possible groups are suggested. Obviously, each organism is unique and different from the others and thus one can recognize five basic sets in the analysis, each containing one kind of organism. Each of these sets can be defined by the attributes shared by all specimens assigned to that kind of organism. But the goal of cladistic analysis is to compare similarities and then to group these individual organisms into larger groups, nested one among the other. The first question to be asked, then, is: What are the general similarities shared among the organisms that could be used to define a group containing all five (termed the *universal set* of the comparison)? There happen to be a number of these similarities: semicircular canals in the head, a chambered heart, visceral (gill) arches at one or more stages of the life cycle, a dorsal nerve cord, a notochord, and appendages of some kind. All these similarities help define the universal set, although only one is sufficient and necessary. All other observed similarities among these organisms, unless they contribute to the definition of the entire group (universal set), must define subgroups.

What are these subgroups? A comparison of similarities begins to yield conflicting results. For example, similarities such as an amniote egg indicate a set comprised of lizard, cat, and mouse. On the other hand, the perch resembles the lamprey in having a non-amniote egg, and thus might be grouped with the latter; and the lizard, perch, and lamprey have neither hair nor mammary glands, implying they might comprise a subgroup.

How are these conflicts to be resolved when the comparison is confined to just these five organisms? It will be recalled that the assumption of evolution provides us with an expectation that groups and subgroups will be characterized by evolutionary novelties. An alternative way of looking at this is to consider evolutionary novelties as changes or modifications of preexistng features (say, from a to a'). Because these preexisting features (a) define a group that also includes the subgroup characterized by the modified feature (a'), these preexisting features are more widely distributed within the universal set. We need, then, to search for more restricted features (novelties) that are interpreted as changed or modified conditions of more widely distributed features. One approach is to compare the morphology of each organism at various stages of its life cycle in order to see if some features change during development, that is, from features that are shared generally with other organisms to those shared by only a few. Perhaps these changes can be used subsequently to hypothesize which features are more widely distributed and which are less so.

At one or more stages of their life cycle all five organisms have much of their skeleton consisting of cartilaginous elements. In the perch, lizard, cat, and mouse, bone is seen to replace much of this cartilage in later developmental stages. These observations suggest the hypothesis that cartilage is a more widely distributed characteristic of skeletons than bone, and that bone, therefore, can be used to characterize a subgroup consisting of the perch, lizard, cat, and mouse. Are there other similarities among these four organisms to corroborate this grouping? All four have jaws, three semicircular canals, paired appendages, and a well-developed vertebral column, to name only a few similarities. If early developmental stages are examined, there are indications that some, or all, of these similarities are modifications of more widespread conditions present in the early stages of all five organisms. For example, the early head skeleton of all five shows many similarities, including the absence of definitive jaws, but the perch, lizard, cat, and mouse go on to develop a lower jaw. The arrangement of three semicircular canals in the perch, lizard, cat, and mouse is a modification of the condition found in early developmental stages of all five organisms (the adult lamprey has two semicircular canals). This evidence seems to support the hypothesis that the perch, lizard, cat, and mouse form a subgroup within a larger group that includes the lamprey.

Within the subgroup just defined, are there further subgroups? The lizard, cat, and mouse have an amniote egg, and this feature can be hypothesized to be a modification of the generalized vertebrate egg, more or less typified by those of the lamprey and perch.

This interpretation is suggested by the fact that all five organisms show similarities in the early stages of development following fertilization, but the lizard, cat, and mouse depart from the general pattern and develop an amniotic membrane.

Finally, among the remaining three organisms, the cat and mouse can be grouped together because of their possession of three ear ossicles, hair, and mammary glands. The three ear ossicles are modified from skeletal elements in the region of the jaw articulation, a pattern common to jawed vertebrates in general; hair and mammary glands can be interpreted, developmentally, as modifications of more general structural characteristics of the vertebrate integument.

By nesting these statements about similarity, the cladogram of figure 2.1 can be constructed. Hypotheses about change in features

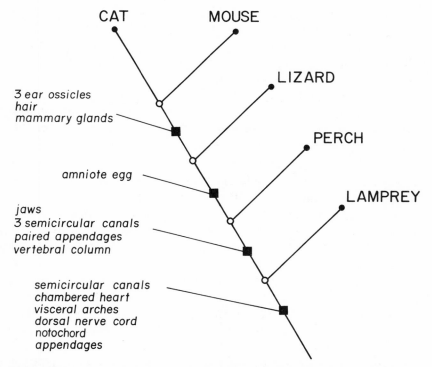

Figure 2.1 A cladogram for five kinds of vertebrates. Each level of the hierarchy (denoted by branch points) is defined by one or more similarities interpreted as evolutionary novelties. (See text.)

(*character transformation*), that is, hypotheses about the hierarchical levels of distribution of the features, were proposed from comparative observations of the developmental stages of the five organisms. This comparative method of constructing nested sets of organisms based on the determination of a feature's distribution, as revealed by developmental sequences, is essentially the method proposed and used by the German anatomist von Baer over 125 years ago. More will be said about the validity and limitations of this method later in the chapter.

An obvious question stemming from an analysis such as the above is: how can a cladogram be constructed in the absence of developmental data? Suppose we have only adults of our five organisms. The key problem is to determine which similarities are evolutionary novelties, and perhaps unexpectedly, the problem is resolved just as in the preceding example by asking how widespread each similarity is. To answer this question other organisms must be taken into consideration.

Our analysis is predicated on constructing a cladogram for the five organisms. Thus, the group as a whole can be characterized by many of the same similarities noted earlier: semicircular canals, dorsal nerve cord, and chambered heart. The problem, as before, is to determine which of the similarities *within* the group are evolutionary novelties. This can be accomplished by determining whether the similarities observed within the group are also found in other organisms outside the group. If they are, then those similarities can be postulated as being too widespread to define subgroups, that is, those similarities are not novelties within the set of five organisms.

Potentially, a vast array of organisms could be compared with the five organisms, so it makes practical sense to place some limits to the kinds that will be studied. Intuitively, it may seem that organisms most like those being investigated would be most appropriate; after all, morphological features within the group must be compared to similar structures in outside organisms. A reasonable first approach, then, is to compare those organisms sharing one or more similarities and see how far the within-group comparison can proceed, i.e., see if a cladogram can be constructed. Thus, several outside groups are similar to the five organisms in having a dorsal hollow nerve cord and a notochord, the lancelet amphioxus and the tunicates (urochordates), for example. Let these groups serve as a basis for postulating novelties within the group of five organisms.

It is observed that the perch, lizard, cat, and mouse are similar in possessing jaws, bone, paired appendages, and three semicircular canals. None of these features is present in the two outside groups, nor do they appear in other organisms that are not at the same time suspected to be part of a larger group defined by the similarities characterizing the five organisms as a whole (chambered heart, semicircular canals). Therefore, we can adopt the hypothesis that the four organisms noted above form a subgroup.

These four organisms now constitute their own group for subsequent analysis, so the outgroups now consist of the lamprey, amphioxus, and tunicates. Within the group of four, what might be a subgroup? The lizard, cat, and mouse have an amniote egg, whereas the perch does not. Upon comparison with the outgroups, the anamniote egg of the perch is seen to be fairly similar to the eggs found in the outgroups. On this basis, it can be postulated that the amniote egg is a novelty defining the lizard, cat, and mouse as a subgroup.

Finally, within these three organisms, is there a subgroup of two? The cat and mouse have three ear ossicles, hair, and mammary glands. None of the outgroups (including now the perch) shares these similarities. Therefore, it can be postulated that the three similarities shared by the cat and mouse are evolutionary novelties defining them as a subgroup.

This approach, called *outgroup comparison,* is similar to the approach using developmental data in that *both are methods used to determine the relative, comparative distribution of observed similarities.* These different levels of similarity form nested patterns which in turn define nested sets of organisms. This comparative method is the general method of cladistic analysis, and the remainder of this chapter will explore the procedure in detail.

Some General Statements about Cladograms

The preceding example can be used to formulate some general statements about cladograms, all of which will be discussed in detail as the chapter unfolds.

First, cladograms are hierarchical in structure because the results of the evolutionary process can be expressed in terms of a branching diagram (figure 2.1). The aspect being expressed in such a diagram is the nested pattern of evolutionary novelties, which themselves define nested sets of groups within groups. This hierarchical structure can also be denoted by nested lists of groups and subgroups as follows (this constitutes a classification; see chapter 5):

> GROUP: Lamprey, perch, lizard, cat, mouse
> SUBGROUP 1a: Lamprey
> SUBGROUP 1b: Perch, lizard, cat, mouse
> SUBGROUP 2a: Perch
> SUBGROUP 2b: Lizard, cat, mouse
> SUBGROUP 3a: Lizard
> SUBGROUP 3b: Cat, mouse
> SUBGROUP 4a: Cat
> SUBGROUP 4b: Mouse

Second, similarities, not differences, define the sets, their nested configuration, and, by extension, the hierarchical levels of the cladogram. Furthermore, from assumptions about the evolutionary process, the only kind of similarity defining groups of organisms is evolutionary novelty. It is the nested pattern of evolutionary novelties that is being expressed by the cladogram and gives it its hierarchical structure.

In practice, if a similarity is perceived, it can be assumed, as a working hypothesis, to be an evolutionary novelty that defines a group of organisms at some hierarchical level. Following the erection of this hypothesis, various kinds of comparative evidence can be used to refute it (e.g., developmental data, outgroup comparison). Thus, in the example given, the possession of jaws was postulated to be an evolutionary novelty defining a group including the perch, lizard, cat, and mouse, whereas the possession of hair was hypothesized to be a novelty defining a subgroup at a lower hierarchical level (cat + mouse).

Given any assemblage of organisms, it is possible to define a universal set and two or more subsets. The existence of a universal set may be taken as axiomatic, given that a worker wishes to construct a cladogram for a specific collection of organisms. On the other hand, most analyses are more open-ended in that the kinds of organisms to be included are not necessarily predetermined. Nevertheless, although not all biologists will necessarily reach agreement about the organisms to be included in any group or subgroup, there must be agreement that these groups within groups can, in principle, be resolved.

Third, for a given universal set of N organisms, it takes at least $N-1$ similarities (novelties) to define the hierarchical structure of any cladogram that might be constructed for those organisms (assuming the cladogram to be fully resolved dichotomously). Thus, one similarity (novelty) is sufficient and necessary to define a group of two or more organisms. In most situations, as in the example, groups will be defined by two or more coincident or congruent similarities, but one is sufficient and necessary (e.g., the amniote egg defining the group lizard + cat + mouse).

The fourth principle is a corollary of the third, namely, a novelty defining a group at one hierarchical level cannot define subgroups at lower levels in the cladogram. The novelty, possession of jaws, defines a group comprising perch + lizard + cat + mouse, but that same similarity cannot define subgroups among these four organisms.

This last principle also implies that the sharing of a postulated novelty by two or more organisms is sufficient to include those organisms within a group defined by that novelty, but the membership of this group cannot be exhaustively determined until the entire membership of the universal set has been compared. After all, the postulated novelty may in fact define a group at a higher level, including perhaps even the universal set. Thus, for any cladogram, it is necessary to specify the highest hierarchical level defined by each hypothesized evolutionary novelty.

Cladistic Analysis: Similarity, Synapomorphy, and Homology

Similarity, Synapomorphy, and Group-forming Procedures

Upon initiating a comparison of organisms, a systematist utilizes perceived similarities to choose those attributes (characters) of the organisms that will then be used for the more detailed comparisons leading to construction of a cladogram. Thus, the choice of characters involves a perception of similarity, i.e., a perception of comparable form and spatial relationships relative to other features of other organisms. In fact, more fundamental, subconscious perceptions probably precede even this elementary level of comparison, and perhaps the basic perceptions are those of "top," "bottom," "anterior," "posterior," and so on. Eventually, at some stage in the thought process, the characters are sufficiently similar to be accepted as the "same" character and are thus worthy of further, more detailed comparison. There is, seemingly, an infinite regress involved in our perception of similarity, and thus perhaps it can be claimed that biologists will compare that which is comparable and will not compare that which is so different—again, in form or spatial relationships to other structures—as to be termed "not the same, and not worthy of further comparison." To a certain extent, it is a contradiction in terms to say that one can compare things that are different; at some level at least, things must be sufficiently similar to invite comparison even though they may eventually be designated "different" and subsequently ignored.

As was previously noted, a major assumption of systematics is that similarities and differences are a consequence of the evolutionary process. Some initial implications of that assumption will now be explored.

Consider two groups of organisms, X and Y, to constitute a universal set and that in both some similarity has been recognized—a *character* is thus identified—but that the similarity is not one of identity. Hence, in X this character can be denoted as *a,* in Y as *a'* (in conventional systematic terminology, *a* and *a'* would be *character-states*).[1] The two ways of specifying relationship—by common ancestry and direct ancestor–descendant (see Chapter 1)—establish the possibility of three evolutionary trees for these two groups (figure 2.2a–c).

How do the character-states of X and Y relate to these evolu-

---

1. The terms "character" and "character-state" merely refer to similarities at two different hierarchical levels. Thus the character "feathers" is a common similarity of all birds, although specific character-states of the character "feathers" (e.g., variation in color, texture, and pattern) would be similarities common to various groups of birds. At the same time, it is apparent that even the character "feathers" could be considered a character-state, say within the vertebrates, if the systematist were considering the "character" to be the vertebrate integument (see also Platnick 1978a).

Thus, in this discussion, the words "character" and "character-state" should be construed to mean relative levels of similarity within a given hierarchy.

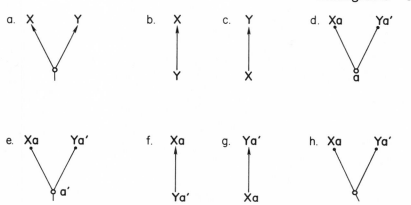

Figure 2.2 (a–c) Three ways the relationships of two groups could be represented by evolutionary trees. (d–g) Character-state *a* is found in X and *a'* in Y, so that there are four character-state trees. (h) The clado-gram for groups X and Y. (See text.)

tionary trees, and what can we infer about the evolutionary history of these character-states? The answer is diagrammed in figure 2.2d–g. Given that there are two groups, X and Y, and two types of relationship, four hypotheses about the evolutionary history of the character-states, *a* and *a'*, can be specified. These four character-state trees can be generalized by converting them to a single clado-gram (figure 2.2h), which is the cladogram describing the information about set-membership contained in the three possible evolutionary trees. (a–c).

Because X and Y are products of some pattern of ancestry and descent, it follows that so too must the character-states *a* and *a'* be a product of that ancestry and descent. The four character-state trees (d–g) indicate only two possibilities, either *a* was transformed into *a'* (d, g), or *a'* was transformed into *a* (e,f). Thus, given two character-states, one of them can be considered *primitive* or *plesiomorphous*—"close to" form (Hennig 1966), and the other *derived* or *apomorphous*—"away from" form. If *a* were primitive and *a'* derived, then only trees 2.2d and 2.2g could be admitted; if *a'* were primitive and *a* derived, then only trees 2.2e and 2.2f would be possible.

Comparisons of two groups such as this are of little significance in systematics because we can make no important generalizations: only one cladogram is possible. However, once the comparison is extended to three groups, important concepts can be introduced.

If three groups are being compared, four cladograms can be

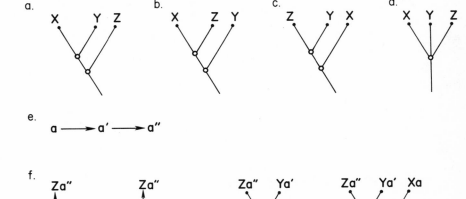

Figure 2.3 (a–d) Four possible cladograms for three groups. (e) A postulated character-state transformation sequence. (f) Four character-state trees for X, Y, and Z assuming the transformation sequence of e. (See text.)

constructed to describe the nested pattern of similarities (figure 2.3a–d). The case in which no resolution is possible (figure 2.3d) will be ignored for the present. Following the example of figure 2.2, let us assume all three groups have a similar, but nonidentical, character such that it can be denoted as character-state a in X, a' in Y, and a'' in Z.

On the basis of this single character it would not appear possible to form a subset within these three groups and thus resolve the universal set (XYZ) into one of the three cladograms (figure 2.3a–c). However, one of the fundamental insights of Hennig (1950, 1966) was to specify a method for forming subsets in situations such as this. Hennig recognized that given some evolutionary change from one character-state, a, to another, a', the investigator may also be able to postulate a further change, from a' to a'' (figure 2.3e). Some simple, well-known examples are the sequences: five toes to four toes to three toes, fully developed eyes to eyes greatly reduced to eyes absent, and undifferentiated epidermis to scales to feathers.

If one has some basis to postulate a transformation from character-state a to a' and subsequently to a'', then there are only four possible trees involving X, Y, and Z upon which the character-state distributions can be mapped (figure 2.3f). The nested structure of all four trees (i.e., the subset, YZ, within the universal set, XYZ) can be represented by a single cladogram (figure 2.3c).

What is the aspect of these trees that enables Y and Z to be placed together in a subset within the cladogram? It is that the ancestor of Z—whether Y or an unspecified common ancestor of both Y and Z—can be postulated to possess a character-state, a', that is *derived* relative to the condition, *a,* present in X or its immediate ancestor (whether unspecified or not). Groups Y and Z form a subset because they share a derived (apomorphous) condition—in this case a' and its modification a''—relative to the primitive (plesiomorphous) condition in X. *The condition of sharing a derived character-state or a later stage in the transformation sequence of a derived character-state is termed synapomorphy.* One of the most important concepts in comparative biology, we shall see that synapomorphy provides the theoretical basis for constructing and testing cladograms.

It should be apparent that within a particular cladogram of three groups perhaps only one of the groups will possess a derived character-state. For example, in figure 2.4 if it were postulated that condition *b* had transformed to *b',* then for this cladogram the derived condition, *b',* would not define a subset. In this example groups X and Y share the postulated primitive condition *b; a shared primitve character-state is termed a symplesiomorphy.* Group Z, on the other hand, is characterized by a derived condition, *b',* not shared with X or Y and such a possession is termed an *autapomorphy* (a'' in group Z would also be an autapomorphy, but *a'* in Y would not, because it is the shared condition, *a',* which is viewed as uniting Z and Y in a subset).

It can be appreciated that neither a symplesiomorphy nor an autapomorphy can be used to combine groups into subsets. Take, for example, the simplest case, in which two groups, X and Y, share a character-state, *a',* whereas a third group, Z, possesses character-state *a* (figure 2.5). If it is postulated that *a* has been transformed to *a',* then X and Y would comprise a subgroup, and the cladogram of figure 2.5a would be indicated. However, if the character-state transformation were in the direction of *a'* to *a,* then this transformation is compatible with any one of the three cladograms (figure 2.5a–c)

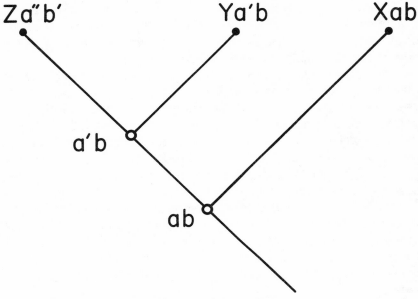

Figure 2.4 In this cladogram, the derived condition *b'* does not define a subset (i.e., Z + Y) but does define Z alone; hence *b'* is termed an *autapomorphy*. The sharing of a primitive condition *b* by Y and X is termed a *symplesiomorphy*.

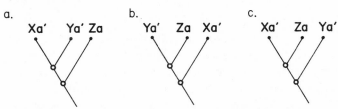

Figure 2.5 A character-state transformation postulated to be from *a'* to *a* would be compatible with any of the cladograms. The conclusion is that symplesiomorphy cannot define subsets.

because character-state *a* would be interpretable as an autapomorphy of Z. In this case, the character defined by the character-states *a* and *a'* contains no information relevant for forming a hypothesis about a particular cladogram, because no synapomorphy is postulated.

Homology and the Concept of Synapomorphy

The concept of homology has been cited frequently as the most important idea in comparative biology. Simpson (1959c:287), for example, calls it "the first and greatest generalization of anatomy." The history of homology is complex (Simpson 1959c), but prior to Darwin, the term, when used, referred basically to similarities among different organisms. After Darwin, it became readily accepted that features which were called homologous obtained their similarity as a consequence of common ancestry. In recent years, both conceptions of homology have been recommended, although for most contemporary systematists the concept of homology implies the idea of common ancestry. For example, Simpson (1961:78) defines homology as "resemblance due to inheritance from a common ancestry." Simpson's view is that homology *is* similarity, but of an inherited kind. Bock (1977:881 and included references) goes further and asserts that similarity should have nothing to do with the definition of homology, rather the definition should be strictly phylogenetic: "Features (or conditions of a feature) in two or more organisms are homologous if they stem phylogenetically from the same feature (or the same condition of the feature) in the immediate common ancestor of these organisms." Bock (1977:882) considers similarity to be the "test" of homology, and by this he presumably means that similarity is the "recognizing criterion" of homology. In claiming that "shared similarity is the *only valid empirical test of homology*" (p. 882, italics in original), Bock virtually adopts the logical position that homology itself is an empirical concept, "knowable" only by observed similarity and not by any phylogenetic criteria. (This assumption will be discussed shortly.)

A second group of systematists, particularly numerical taxonomists, finds the phylogenetic definition of homology to be unsatisfactory. These definitions are, they believe, self-defeating because "it is the primary purpose of the phylogenetic school of systematics [here they are really referring to the school of evolutionary systematics] to work out phylogenies, and for this they need homologies that are not defined in terms of the conclusions they wish to reach" (Sneath and Sokal 1973:75). Sneath and Sokal have identified aspects of this definition that bother systematists concerned with methods to reconstruct phylogeny. They propose to return to an "operational definition": "When we say that two characters are operationally homologous, we imply that they are very much alike in general and in particular" (p. 79). This definition, it would seem, is remarkably similar to Bock's "test" for homology.

The diversity of opinion noted above is testimony to the fact that homology continues to occupy an important position in discussions about comparative theory and methodology. However, most of this discourse has contributed only marginally to the important methodological question of comparative biology: how do we reconstruct the history of life? On the one hand, the definitions of Simpson and Bock do not contain intrinsically any implications for methodological generalizations, and as such they create a paradox, noted by Sneath and Sokal. Are "homologies" to be used to reconstruct phylogeny or must we, as the definitions of Simpson and Bock imply, have an established phylogeny before identifying homologies? If the former, then their definition would seem superfluous; if the latter, then the homology concept would seem to be of little consequence for systematics. On the other hand, the operational approach of Sneath and Sokal is also unsuitable. Almost universally, homology has meant something more to most systematists than mere similarity; furthermore, the Sokal and Sneath definition implies little about the goal of comparative biology, the reconstruction of the history of life.

These opposing viewpoints constitute what might be called the "problem of homology," and neither the schools of evolutionary systematics nor numerical taxonomy have offered a conceptual solution to this problem. The solution does not lie in proposing a concise, easily remembered definition of the word "homology," but rather in understanding the fairly simple relationship between inherited similarity on the one hand and the use of comparative analysis to discover group-membership on the other: the solution to the "homology problem" is the concept of synapomorphy.

Homologous similarities are inferred inherited similarities that define subsets of organisms at some hierarchical level within a universal set of organisms. Viewed in this way, homology can be conceptualized simply as synapomorphy (including symplesiomorphy; see below). Indeed, there can be little doubt that many systematists, particularly those who have formed the post-Darwinian tradition in comparative biology, have sensed the connection between homology and synapomorphy, although it is only recently that this concept has been made explicit (see, for example, Hennig 1966; Nelson 1970; Wiley 1975; Cracraft 1978; Gaffney 1979).

Synapomorphy subsumes the concept of symplesiomorphy. Synapomorphies are shared similarities (homologies) inherited from an immediate common ancestor; symplesiomorphies are shared similarities (homologies) inherited from ancestors more remote than the immediate common ancestor. But a symplesiomorphy can be viewed as a similarity whose level of synapomorphy is unresolved; as more groups are added to the comparison, what was once considered a symplesiomorphy becomes a synapomorphy defining a set of organisms at a higher hierarchical level. The question of importance is: at what level of the hierarchy does a shared similarity define a set? For example, the similarity of dense body hair cannot define a set including a mouse and a lion but excluding a human, because the shared possession of dense body hair between the mouse and lion is, at that hierarchical level, a symplesiomorphy, that is, there are other organisms with dense body hair excluded from the set. As the hierarchical level is increased by adding other groups with hair, this similarity is eventually seen to define a set biologists call Mammalia, and at this level hair is a shared derived similarity, a synapomorphy.

The concepts of synapomorphy and homology can be viewed in another way, one which pertains to defining group membership. As will be developed in more detail later, the "test" for synapomorphy is not empirically observed similarity (as was Bock's test for homology, noted above), but the congruence of other hypothesized synapomorphies in defining sets of a cladogram and, ultimately, the systematist's preference for that cladogram in contrast to alternatives that might have been chosen. For example, given the cladogram shown in figure 2.6a, the postulated synapomorphy a' defines a subset AB within the set ABC. However, if from other evidence the investigator had reason to prefer the cladogram of figure 2.6b, then the postulated "synapomorphy" does not define a set within the cladogram. Is the similarity a', therefore, a synapomorphy? No, not in the sense that a' defines a subset within the hierarchical level ABC. Then, how can we explain the distribution of the similarity a' in A and B if there is reason to prefer the cladogram of figure 2.6b? There are two explanations. First, the similarity a' may be a synapomorphy, but one defining set ABC or some still larger set. If this were the case, condi-

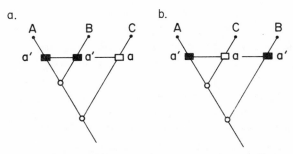

Figure 2.6 (a) Character-state a' seems to define a set A + B and thus might be postulated to be a synapomorphy. (b) If there is sufficient evidence to accept this cladogram, then a' can no longer be a synapomorphy at this hierarchical level but must either be a primitive similarity (symplesiomorphy) or a convergence.

tion a' in A and B would be a shared primitive similarity, and the condition a in C would be derived, an autapomorphy. In order to determine this situation methodologically, the analysis must be extended to include groups of organisms other than A, B, and C, and also to include other features. Secondly, if for cladogram 2.6b evidence could be presented that condition a is indeed primitive and a' derived (for example, through comparison with other groups of organisms), then similarity a' would have to be interpreted as having been derived independently. In this case, a' would be a *convergent,* or *nonhomologous,* similarity.

This example, then, implies that similarities may be considered homologous (synapomorphous, symplesiomorphous) or nonhomologous (convergent) depending on their ability to define subsets within a particular cladogram. If the point of reference (a cladogram) were to change, our view about what is or is not a homology would be similarly altered. The notion that the concepts of "homology" and "nonhomology" are not empirically determined and not factual descriptions of observed similarity, but rather a consequence of set-definition, clarifies much of the confusion in the literature on homology. Even if evolutionary connotations are imparted to the idea of homology, properties of set-definition continue to be essential criteria for applying the concept. This should not be too surprising, since the resolution of nested subsets is the primary concern of cladistic analysis, and of comparative biology in general.

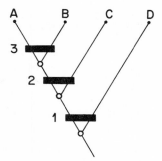

Figure 2.7 A cladogram of four groups. A is the sister-group of B, A + B the sister-group of C, and A + B + C the sister-group of D. Black rectangles signify defining synapomorphies.

Two additional concepts are associated with the idea that synapomorphies define nested sets within a cladogram. First, the coordinate subsets of any set defined by a synapomorphy are termed *sister-groups*. Therefore, in figure 2.7, group A is the sister-group of B, group A + B the sister-group of C, and group A + B + C the sister-group of D. The concept of sister-groups is relative and not absolute and, as we shall see, is itself also a hypothesis to be tested. Its application in any specific instance depends upon the number of taxa included in the comparison (universal set) and the perceived pattern of synapomorphies. In figure 2.7, if B had been omitted for some reason, then groups A and C would be sister-groups united by synapomorphy 2.

A second concept is to some extent one of classification (chapter 5) but it has its basis in set-definition and synapomorphy. A set of organisms is said to be *monophyletic* if it includes the stem species (if known) and *all* those kinds of organisms hypothesized to have descended from it. It follows that the concept of monophyly is applicable only to groups of two or more species (see Chapter 3). An important corollary of this definition is that monophyletic groups are defined on the basis of one or more synapomorphies. The sets AB, ABC, and ABCD of figure 2.7 are monophyletic, but the set AC is *nonmonophyletic* because B has been excluded even though a synapomorphy indicates its inclusion. Sets of organisms may have names and be part of the Linnaean hierarchy, in which case mono-phyly comes within the purview of classification theory. (This will be discussed more fully in chapter 5.)

The Search for Nested Sets

The key methodological problem in comparative biology might be viewed as a search for synapomorphy. More specifically, the search is for nested patterns of synapomorphy, and cladograms are hypotheses about those nested patterns.

Cladograms, and thus nested synapomorphy statements, have a property in common with all scientific hypotheses: that of prediction (see Platnick 1978b). The prediction is that synapomorphies will be congruent with one another, and this prediction is based on our ex-pectation that the evolutionary process has produced a single, unique history of life, and that changes in organisms and their fea-tures will conform to the pattern of this history more often than not. The essence of cladistic analysis, therefore, is the formulation of hy-potheses of synapomorphy and their nested pattern. Subsequent evaluation of hypotheses of synapomorphy and of the cladogram it-self rests with the prediction that future synapomorphy statements will continue to be congruent with those of the cladogram. As we shall see, this implies a preference for the cladogram which max-imizes the congruence of individual synapomorphy statements. This view of cladistic analysis was implicit in the example of figure 2.1:

Lamprey, perch, lizard, cat, mouse: dorsal nerve cord
Perch, lizard, cat, mouse: jaws
Lizard, cat, mouse: amniote egg
Cat, mouse: hair

There is the expectation that future synapomorphy statements (notochord, semicircular canals, paired appendages, three ear os-sicles, and so on) will conform to this nested pattern. If they do, then our confidence grows in this cladogram as a reliable representation of the pattern of life's history. If they do not, then perhaps an alterna-tive cladogram will represent that pattern better.

Cladistic Analysis: Taxa and Characters

In this section, the practical problem of constructing cladograms will be considered. Three broad topics are discussed: the kinds of groups of organisms that are subjected to a cladistic analysis, the kinds of characters that provide the basic data for the analysis, and, finally, the procedures that might be followed in formulating and testing hypotheses of synapomorphy patterns (the cladogram itself).

The remainder of this chapter is a "manual" on cladistic analysis only in a very restricted sense. No one description of methods can lead investigators step by step through the countless problems and decisions which arise in a real biological example. Our purpose, rather, is to provide the conceptual foundation for the investigator's own efforts, and, by using examples, we trust the conceptual foundation will be brought closer to the real world situations confronted by practical experience. Our own experiences, and those of our students, have convinced us the only way to fully understand the conceptual aspects of cladistic analysis is to apply the methodological theory to a group of organisms. It is a process of trial and error, and it has rewards beyond the success of a finished cladogram: applying the methodology can also commit one to further understanding and clarification of the underlying concepts.

The Elements of Cladograms: Kinds of Taxonomic Units

In the earlier biological example (figure 2.1) the groups of the sets were loosely termed "organisms." It seems doubtful this usage was confusing, since to most, if not all, biologists the five "organisms" would be recognized to be of a distinct kind, what are typically called "species." Before proceeding to the details of cladistic analysis, the exact nature of these biological groups should be discussed.

The fundamental units of comparison in systematic studies are *taxa* (singular, *taxon*). Taxa are defined as collections (sets) of individual organisms that are sufficiently distinct from other sets to be given formal names and placed in the Linnaean hierarchy (see Mayr 1969:4–5; Simpson 1961:19). For example, the taxon name *Columba* refers to a genus (a category of the Linnaean hierarchy) of pigeons, the taxon name *Columba livia* to a species of pigeon, and the taxon name Columbidae to the family including all pigeons. The lowest hierarchical category within which formal taxa are sometimes classified is the subspecies, thus *Columba livia livia* (see chapter 5).

If systematists were asked to name the primary taxonomic level at which systematic analyses are undertaken, most would answer that of the species. How one might identify a collection of individuals as belonging to a species or whether a precise delimitation of species is in fact necessary for systematic work to proceed are questions that have been vigorously debated (Mayr 1963, 1969; Sokal and Crovello 1970; Sneath and Sokal 1973; Wiley 1978). We discuss these problems in chapter 3. It is important to note that, although species are commonly defined in terms of reproductive criteria, relatively few practicing systematists directly use these criteria when applying the species concept in systematic studies (see chapter 3).

Our immediate task at this point is to specify the kinds of taxa systematists might want to investigate using cladistic analysis. The one basic assumption is that each included taxon composed of two or more species is strictly monophyletic.[2] In practical terms this means the systematist has some reason to accept a corroborated hypothesis of their monophyly, and the reason, as we have noted previously, is evidence of congruent synapomorphies. A hypothesis of monophyly for any supraspecific taxon must be based on synapomorphous similarity. In general, most species-level taxa will also be defined in terms of synapomorphy, but, as will be discussed in chapter 3 there may be cases in which this need not be a requirement. Cladistic analysis can be performed, therefore, using taxa of species-level rank or higher.

---

2. The concept of monophyly is applicable only to groups of two or more species. Obviously, when speaking of single species, the concept of monophyly is inappropriate since stem (ancestral) species and descendant species are necessarily excluded from the discussion.

This observation raises a more general point relevant at this juncture: the ancestral species within any monophyletic group, if present in a sample and correctly analyzed, will be diagrammed as a monotypic sister-group of the remainder of the species contained in that group (see chapter 5). If the ancestral species is raised to some higher rank (see chapter 5) for classificatory purposes, e.g., placed in a separate family, then that family will not be monophyletic. In general, taxa of any rank which consist solely of a single species ancestral to other species within a monophyletic group cannot themselves be monophyletic. Every monophyletic group, under Hennig's definition, will contain, as a minimum, one such example (i.e., a stem species). As a corollary, such nonmonophyletic taxa will not possess any uniquely derived characters (autapomorphies).

Similarity and Characters

Comparison among taxa implies the search for similarity, and cladograms are general expressions of the patterns of those similarities.
This statement forms the conceptual basis for understanding the process of constructing and testing cladograms: that process is merely
the systematist's method of arriving at general statements about similarity. In one sense, the characters of organisms are the carriers of
similarity. This view emphasizes an important point: it is not the characters themselves that are of significance, but rather the similarities.
Something more than a semantic issue is involved here. There has
been considerable discourse within systematics about characters
and character-states, reliable versus unreliable characters, the
"weight" of characters, and so on, but little of this literature mentions
or emphasizes similarity. These discussions of characters have dis-
tanced themselves from the central concept of comparison: similarity.
The ensuing pages, therefore, focus on similarity and its role in comparison and the construction of cladograms.

Some general statements about similarity were introduced earlier in the chapter. All comparison—biological or not—involves the
correspondence of similarities. Arms could be compared with
heads, yet the basis for that comparison still remains a search for
similarity; thus, both arms and heads might be seen to be comparable in having similar tissues, cells, chemical characteristics, and so
forth. This comparison would resolve itself to comparing not the
characters called "arms" and "heads," but rather the characters
called "tissues," "cells," and "chemical characteristics." However,
similarities must be named, and it is these names that we call "attributes" or "characters." If the process of comparison has some general significance, it is that as soon as a character is designated, then
so too is a similarity: in comparative biology the words are conceptual, if not entirely operational, synonyms—a character is a name for
a similarity. Thus, "paired appendages" may refer to a similarity that
is common to a set of vertebrates, "differentiated paired appendages" to a similarity of a more restrictive subset, and a "pentadactyl foot" to a similarity of a still more restrictive subset. This suggests
that discussions in the systematic literature about characters and
character-states are really about similarity. A "pentadactyl appen-
dage" as a character, and "four-toed" or "three-toed" conditions as

character-states would seem to have little intrinsic usefulness in comparison if they did not refer to the similarities implied by those names. A further implication is that the similarities are defining sets and subsets. The implication itself is fascinating—name an attribute, any attribute, and the working hypothesis is that it is a similarity defining a set. The set may not be "natural" in an evolutionary sense ("eyes" of vertebrates grouped with "eyes" of arthropods) but, as will be pointed out in what follows, problems of this sort are resolved by comparing "characters" defining higher hierarchical levels—the essence of testing cladograms. In any case, more restrictive "characters" (read "similarities") can be seen to define nested sets.

What kinds of similarities observed in taxa can be used in cladistic analysis? The answer is that all observed similarities may be of interest and potentially useful. Logically, those similarities which are *intrinsic* to the species comprising the taxa would seem to be more immediately useful for cladistic analysis than those which are *extrinsic* to the species. By intrinsic, we mean similarities of anatomy, physiology, biochemistry, behavior, genetics, and development; by extrinsic, we mean the similarities among species that might be observed in their distribution in space and time. (Intrinsic and extrinsic "properties" are also treated in the next chapter on species.)

Anatomical similarities are clearly the most commonly used similarities in biological comparison. Some of these were illustrated in the cladogram at the beginning of the chapter (figure 2.1); more will be included in later examples. Anatomical similarities predominate in systematic analysis primarily because they are the most easily observed, and most can be studied in preserved organisms. Nonanatomical similarities, while important in specific cases, have been less widely adopted in systematics owing generally to a lack of comparative data. Many nonanatomical similarities must be observed in living organisms or under experimental conditions requiring very elaborate and expensive techniques. Nevertheless, through the years biologists have learned a great deal about the comparative aspects of these kinds of similarities. An example of one of the more extensive attempts at using nonanatomical similarities to construct and test cladograms is Løvtrup's (1977) analysis of the origin and evolution of the vertebrates.

Physiological similarities have seldom been adopted in systematic studies, owing mainly to a lack of comparative information. Perhaps the best summary of physiological similarities within animals can be found in textbooks of comparative physiology (e.g., Prosser and Brown 1965). A few examples will suffice to illustrate the use of physiological similarities in defining groups of organisms.

Almost all animals (and some plants) possess respiratory pigments that bind oxygen and transport it from a respiratory surface to other body tissues. In most organisms this pigment is hemoglobin, but other kinds of pigments also exist. One of these, chlorocruorin, is found only in the plasma of species in four families of polychaete worms, Sabellidae, Serpulidae, Chlorhaemidae, and Ampharetidae (Prosser and Brown 1965). Although some species of Serpulidae also have hemoglobin, shared possession of chlorocruorin defines a group including the four families.

Several interesting similarities defining groups of organisms are associated with the physiology of digestion. Virtually all animals studied to date possess amylases for the digestion of starches, and the possession of this physiological property defines a group including all animal life. Shared similarities of other digestive enzymes also define groups: pepsin, for example, is found only in the vertebrates (Prosser and Brown 1965).

Physiological similarities will undoubtedly be of great value to systematists in the future. Several difficulties will always accompany their use, but by no means are these difficulties restricted to physiological similarities. First, compared to anatomical similarities, relatively little is known about the occurrence of any given similarity throughout the taxon under study. Second, statements that a certain physiological process is or is not present or whether it is similar among taxa are often made only on the basis of functional observations, and detailed studies of the precise composition of chemical products or the biochemical pathways involved in their elaboration frequently have not been undertaken. Thus it is often problematical that the "same" or "similar" physiological processes are being observed or compared.

Biochemical similarities, to the extent that they can be distinguished from physiological similarities, have proven helpful to systematists in defining the groups of cladograms. Løvtrup (1977), for example, lists a large number of biochemical similarities of molluscs and vertebrates said to define a group distinct from other animals:

these similarities include the distribution of specific glycosaminogly-
cans (hyaluronate, chondroitin, and keratin sulfate), epidermal structural proteins (epidermin), and cartilage (also found in protochordate
groups), among others.

With the exception of the findings of comparative anatomy, no
other kind of similarity has been utilized by systematists as much as
that of behavior. Most behavioral similarities have been observed at
the species level, and the emphasis most often has been on the
demonstration of how one species is distinct from another in behavioral characteristics. Expressed somewhat differently, and perhaps
more correctly, these investigations use behavioral similarity to define groups of individual organisms which are then called species.

Behavioral similarities can also be used to form groups of supraspecific taxa. In his extensive comparative studies of stork behavior, Kahl (1971, 1972) discovered a series of behavioral similarities defining a subset within the stork family, Ciconiidae, that
includes the four species of wood storks (*Mycteria americana, Ibis
ibis, I. leucocephalus,* and *I. cinereus*). These four species show
similarities in several stereotyped behavioral patterns termed display preening, balancing posture, and gaping (figure 2.8). Kahl was
unable to find behavioral similarities defining subsets within the four
species of wood storks.

Gorman (1968) has employed behavioral similarities in the display behavior of the species in the lizard genus *Anolis* to define subsets within the genus. The *roquet* species-group, consisting of six
species, is defined by the possession of a shared similarity, a dis-

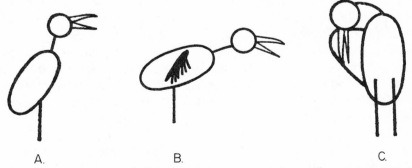

A.   B.   C.

Figure 2.8 Behavioral similarities in display patterns used to define wood
storks: (a) gaping; (b) balancing posture; (c) display preening. (From Kahl
1972, by permission of the British Ornithologists' Union.)

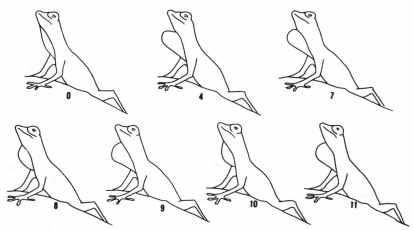

Figure 2.9 A behavioral similarity in dewlap display in *Anolis* lizards used to define various species-groups. (From Gorman 1968.)

play in which the dewlap (a fold of skin trom the throat) is extended and then held in this position for the entire display; other species of *Anolis* rapidly pump the dewlap in and out during the display (figure 2.9).

An increasing number of systematists are using "genetic" or "molecular" data to form sets of taxa (see Ayala 1976, and Dobzhansky et al. 1977, for an introduction to the types of work being undertaken). These investigations include the analysis of karyotypes (chromosome number and structure), indirect techniques of estimating "genetic distances" among organisms (DNA hybridization, electrophoresis, and immunological similarity), or the "direct" comparison of the amino acid sequences of a given protein among various taxa (even this may be considered an "indirect" genetic technique, in the sense that the results only provide an estimate of the similarity in nucleotide sequences).

Most of the studies on karyotypes, chromosome structure, and electrophoretic mobility are directed at forming groups of organisms at the species level. For example, Greenbaum and Baker (1976) analyzed the chromosome number in populations of two mainland species of bats of the genus *Macrotus*. One species, *M. californicus,* has a diploid chromosome number of 40, whereas the other species, *M. waterhousii,* has a diploid number of 46. Individuals in populations on the islands of Jamaica and Haiti were found to have a diploid number of 46 and therefore were included in the species *M. waterhousii*.[3] This is only one example of many that could be given to illustrate the use of karyotypes to define species-level taxa.

Techniques such as DNA hybridization, starch-gel electrophoresis, or immunological reactions can be used to estimate the degree of genetic similarity among species. The methods of analysis adopted in these studies, although routinely used to produce branching diagrams depicting concepts of relationship, are qualitatively different from the methods discussed in this book, in that they make no attempt to distinguish between primitive and derived similarities. Instead, similarity (or dissimilarity) matrices are produced; for example, a matrix might show the degree to which species cross react immunologically with one another. These numerical data matrices are then analyzed by various clustering techniques to yield branching diagrams. This approach is analogous methodologically to the concept of general "overall similarity" adopted by numerical taxonomy in that similarity is expressed as a numerical value in a data matrix, and then the latter is used to produce a branching diagram of some kind. We further discuss numerical taxonomic techniques in chapter 5; we merely note here that cladistic analysis as defined and described in this book is theoretically and methodologically distinct from the techniques mentioned above.

In recent years, biochemists have determined amino acid sequences for homologous proteins in different taxa, and comparison of these sequences has permitted cladistic hypotheses to be proposed. Unlike the numerical "similarity" measures yielded by DNA hybridization or immunological reaction techniques, comparisons among amino acid sequences can be evaluated in terms of primitive and derived similarities. For example, in cytochrome *c* proteins humans and rhesus monkeys differ by only one amino acid: at position 66, the human has isoleucine and the rhesus monkey has threonine. Examination of nonprimate mammals indicates that threonine is widespread. Therefore, it can be postulated that, within primates, at position 66 threonine is primitive and isoleucine derived. Crowson (1972) has discussed in more detail a procedure for comparing amino acid sequences by this approach.

3. For this to be a valid procedure, of course, it would have to be shown that a chromosome number of 46 is actually a derived (synapomorphous) condition, otherwise the Jamaican population might not necessarily belong to *M. waterhousii*.

Table 2.1 Developmental Similarities Used to Define the Two Currently Accepted Animal Superphyla[a]

Annelid superphylum	Echinoderm superphylum
Spiral cleavage	Radial cleavage
Blastopore = mouth	Blastopore = anus
Schizocoelic coelom	Enterocoelic coelom
Determinate cleavage	Indeterminate cleavage
Delaminates nervous system	Invaginates nervous system
Ectodermal skeleton	Mesodermal skeleton
Trochophore larva	Pluteus-type larva

[a] From Løvtrup (1977:60), after Kerkut (1960).

The final kind of intrinsic similarity to be illustrated is that of development. Similarities in development have played an important role, historically and scientifically, in the definition of the major groups of animals. Kerkut (1960), for example, discusses many developmental similarities indicating a major division of the animal kingdom into the annelid superphylum and the echinoderm superphylum; the similarities defining these two postulated sets are shown in table 2.1. Although Løvtrup (1977) has questioned the interpretation of these observed similarities, and thus the definition of the groups themselves, the similarities exemplify the kinds of evidence to be found in developmental patterns. The application of developmental similarities is discussed in more detail later in the chapter.

To many practicing systematists, intrinsic similarities are viewed as a direct manifestation of an organism's genotype. To be sure, the expression of most of these similarities, perhaps to be equated with the concept of phenotype, may well be under genetic control. The point is, however, that with very few exceptions systematists do not have knowledge about the genetics (or developmental expression of these gene products) of the similarities they are comparing, nor does this information appear to be crucial. Cladistic analysis functions to discriminate pattern—presumably the consequence of an evolutionary process—not the underlying causal mechanisms of that pattern.

Extrinsic features of taxa are those associated with their distribution in space and time. Although extrinsic similarities among taxa are certainly of interest to systematists, the extent to which they bear upon cladistic analysis is slight. Cladograms are hypotheses about the distribution of intrinsic properties; extrinsic features lack a 1:1 correlation with nested sets of taxa. It would appear, therefore, that extrinsic similarities are by nature different from intrinsic similarities. Only intrinsic features are discussed further with respect to the construction of cladograms.

Cladistic Analysis: Hypotheses and Their Evaluation

There are two major components to cladistic analysis. First is the formulation of a hypothesis about nested sets of taxa, and second is the evaluation of this hypothesis relative to alternative hypotheses that might be considered. These two components are those of empirical science in general; one begins with a question or a problem, which may suggest an array of possible solutions (hypotheses) that must in turn be evaluated (tested).

The primary question of systematics is: what has been the history of life? The view developed in this chapter is that cladograms are hypotheses about the pattern of that history, and statements about synapomorphy provide the basis for evaluating alternative cladograms. In what follows we will focus on three issues: the development of cladistic hypotheses, the analysis of synapomorphy, and the evaluation of cladistic hypotheses.

Development of Cladistic Hypotheses (Cladograms)

A cladistic hypothesis is a cladogram specifying a pattern of relationships (nested sets) among taxa that is a consequence of a nested pattern of synapomorphy. It is not necessary, of course, to have knowledge of synapomorphies or their possible nested pattern prior to proposing a cladogram, because the hypothesis stands on its own and is eventually subject to rejection or corroboration. Put another way, a cladistic hypothesis (cladogram) is a hypothesis of monophyly (see Gaffney 1979). In its simplest form it is a statement that says: "taxon A is more closely related to taxon B than either is to taxon C" (figure 2.10a). This statement, called a *three-taxon statement,* can be variously interpreted. In much of the literature on

a.   b.   c.

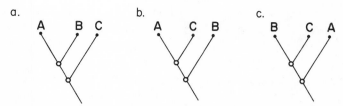

Figure 2.10 Three alternative three-taxon statement cladograms. (See text.)

cladistic methodology, it has meant: "taxon A and taxon B shared a
more recent common ancestor than either did with taxon C," and in
this sense the statement is one of phylogenetic relationship *sensu*
Hennig (1966). The ensuing discussion will admit this interpretation,
but something more specific will also be implied: taxa A and B share
a pattern of synapomorphy unique to themselves and nested within a
more encompassing synapomorphy pattern defining taxa A, B, and
C. Hence in this sense *relationship* ultimately refers to a hypothesis of nested synapomorphy.

The hypothesis stated above, that taxon A is more closely related to taxon B than either is to taxon C (figure 2.10a), can be
proposed in the absence of any supporting evidence, for it is implied in the formulation of such a cladogram that this hypothesis
stands in competition with alternative hypotheses. In the case of
three taxa, there are two completely dichotomous alternatives (figure
2.10b,c), and it is the purpose of cladistic analysis to provide a
method to chose among the three hypotheses.

How might a cladistic (cladogram) hypothesis be formulated?
What taxa are to be included? It can be assumed at this point that a
systematist undertaking a cladistic analysis will have a particular
problem in mind with respect to a specific group of organisms. Initially, the taxa will be those delineated in prior systematic treatments
of the group. It is important that all taxa included in a cladistic analysis be monophyletic, i.e., be definable by synapomorphous features
of their own. Thus recognition of these taxa is equivalent to formulating a series of hypotheses of monophyly. In most instances, systematic tradition will have accepted the monophyly of the basic taxa, in
which case the next concern is to develop a hypothesis of relationships for them. If, on the other hand, this systematic tradition has
indicated uncertainty about the monophyly of one or more of these

taxa, it will be necessary to investigate this problem further before attempting to group these taxa into larger groups. In either case, the basic taxa of a cladistic hypothesis can be viewed as a series of lower-level hypotheses, themselves subject to testing.

Once it has been accepted that certain basic taxa will form the subject matter for a subsequent cladistic analysis, the next step is the formulation of alternative hypotheses that can be tested. Although there are no established formal rules for erecting these hypotheses, systematists have traditionally followed only a few methods.

1. Hypotheses suggested by prior systematic work To a greater or lesser extent, all systematic analyses rely on the hypotheses of prior systematic work. Indeed, in those groups previously receiving considerable attention, most, if not all, of the possible hypotheses may already have been proposed and discussed. Although many of these investigations may not have been undertaken from a cladistic viewpoint, the hypotheses generated can often serve as a starting point for a cladistic analysis nonetheless. For example, in his cladistic study of the interrelationships of several major groups of actinopterygian fishes, Wiley (1976) identified four hypotheses of relationships that had been previously suggested, and then proceeded to evaluate each (his study will be discussed further later in the chapter).

Unless one is investigating a poorly studied group or attempting to decipher the relationships of newly described taxa, prior systematic work will probably be a major source of hypotheses to be tested. Even so, there are at least three alternative methods for generating these hypotheses.

2. Hypotheses generated by enumeration Given a specific number of taxa to be analyzed, the systematist might list all the cladograms possible for those taxa and then subject each hypothesis to testing. If one were concerned with a very small number of taxa (say, three or four), then this procedure might prove feasible. But for 5 taxa 32 fully resolved cladograms are possible, and the number increases markedly thereafter (e.g., there are 10,395 possible cladograms for 7 taxa). Clearly, enumeration will not be an important method for generating hypotheses unless more complex situations are resolved to statements of the three-taxon type (figure 2.10).

3. Hypotheses suggested by general similarity Many clusters of taxa are suggested by aspects of general similarity. Historically, general similarity has been the principle means used by systematists to group taxa in terms of relationships and to form classifications. Because general similarity includes a component of symplesiomorphy, it cannot be used, in itself, to define monophyletic groups. Nonetheless, general similarity sometimes can lead to the formation of hypotheses that can later be tested by synapomorphy. If fact, as is often the case, an observed similarity in two or more taxa can serve as a preliminary reason for hypothesizing monophyly, and then the similarity can be compared more broadly in other taxa in order to assess whether it is a primitive or derived condition (a method that does not differ materially from the next approach).

4. Hypotheses of synapomorphy In many cladistic analyses, hypotheses of synapomorphy are used directly to formulate hypotheses of monophyly. Those latter hypotheses are in turn tested by additional synapomorphies that might be postulated. In following this procedure, the line between hypothesis formulation and hypothesis testing becomes blurred. There is usually a prior acceptance of basic taxa, similarities are compared, and, on the basis of one or more criteria, synapomorphies are postulated. These synapomorphies lead directly to one or more phylogenetic hypotheses. Examples of this approach will be presented shortly.

Analysis of Synapomorphy

Without doubt, the most important part of any cladistic study is the analysis of synapomorphy. Not only is the delimitation of the basic taxa dependent upon an assessment of synapomorphy but so too are the eventual interrelationships determined for those taxa. Hence, there is ample reason for so many systematists to have expended considerable time and thought to this subject. Most of this literature falls under the rubric of *character analysis,* and many workers have provided "criteria," "rules," and so forth, to recognize primitive-derived conditions of similarities. But if the rationale for cladistic analysis is accepted, that is, if it is agreed that cladograms are expressions of nested patterns of synapomorphy, then the problem of character analysis is one of resolving character distributions. No matter what kinds of similarities are used—anatomical, physiolog-

ical, behavioral, biochemical, developmental, or genetic—
synapomorphy patterns are resolved by nesting more restricted
statements of similarity within more general ones.

The analysis of synapomorphy begins with the recognition of a
similarity—a postulated homology—and its methodological designa-
tion as a character. Concomitant with this is the recognition that such
a character exhibits some variability in terms of discrete character-
states distributed in some pattern among the basic taxa. Some
writers have suggested that these character-states can be arranged
in a *morphocline,* a sequence presumed to reflect the probable
pathway (not direction) of change among the character-states (Maslin 1952; Schaeffer, Hecht, and Eldredge 1972; Ross 1974). The relevant problem thus becomes the establishment of the morphocline's
polarity, i.e., the determination of the primitive and derived ends of
the morphocline. For example, given the observation that fossil and
Recent horses have two toes, three toes, and one toe, the morphocline one-toed—two-toed—three-toed might be established as
most reasonable. Subsequent analysis might suggest the polarity to
be three-toed → two-toed → one-toed, although if one ignores the
idea of a morphocline other possibilities clearly can be considered
(figure 2.11). (It also could be suggested that the three character-
states might have been derived from some other, unknown
character-state, but such an assumption would be ad hoc.)

Perhaps once character-states are delimited, the perception of a
morphocline may be inevitable in the thought process. However, the
establishment of a morphocline may be more heuristic than necessary, and if all possible polarity sequences are not considered, it

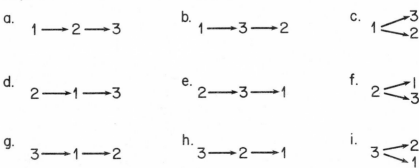

Figure 2.11 Possible character-state-transformation sequences for the three
character-states (1) one-toed, (2) two-toed, and (3) three-toed.

may sometimes be difficult to guard against a preconceived notion of polarity (see also Gaffney 1979). It seems best to avoid unnecessary assumptions by simply identifying character-states and then following procedures to postulate which are relatively primitive and which relatively derived. To repeat: "characters" and "character-states" merely signify different hierarchical levels of synapomorphous similarity.

The concept of nested sets of synapomorphy makes sense only if it is assumed that change has taken place, that somehow a group of organisms, itself a subset of a larger set of organisms, is characterized by a change (a transformation) in one or more features that can be said to characterize the group as a whole. Thus, in analyzing synapomorphy, the problem is to develop methods to postulate the sequence of change. Basically, systematists have recognized three ways to identify a transformation sequence: (a) the application of paleontology, (b) the application of ontogeny, and (c) the application of outgroup comparison. We will now discuss each of these in order.

1. The Application of Paleontology A traditional assumption of paleontological practice has been that fossil taxa are the key to the history of life, and that, in their absence, all knowledge about relationships is speculative at best. Not only has this viewpoint concerned itself with the identification of possible ancestral-descendant sequences of taxa but also with the determination of primitive-derived sequences of character transformations. Indeed, this paleontological viewpoint can be characterized quite simply: knowledge of phylogeny is obtainable only through direct empirical observation of the historical record. Among the recent pronouncements on this subject is the following passage from Gingerich and Schoeninger (1977:488), who speak for a tradition extending back well into the nineteenth century:

> These basic ideas about phyletic evolution, like the basic principles of mechanics, could not be predicted from theory alone—they are essentially empirical, and it is only with an adequate fossil record that any real understanding can be gained of the importance of size increase, irreversibility, or parallelism. . . . There is, in the absence of an adequate fossil record, no way to be certain which characteristics of an organism are primitive and which are derived, or which evolved independently in different lineages. Knowledge of both is essential to reconstruct a phylogeny on the basis of living forms alone.

Other paleontologists, to be sure, do not necessarily express themselves in as positive a manner as Gingerich and Schoeninger. Nevertheless, within paleontology, data from fossils are generally considered to be useful in determining primitive-derived sequences. Typically, this view is expressed as follows:

> For groups with an extensive fossil record, the character state first appearing in the record is likely the ancestral state. (Harper 1976:185)

> One should have sound biological reasons for hypothesizing a character state to be primitive when it appears late in the fossil record, and when at the same time taxa with what are identified as advanced states abound in earlier strata. . . . It is therefore expected that the primitive, rather than advanced, character states should occur with greater frequency at earlier dates. (Szalay 1977:17)

The general notion that the fossil record can tell us which character-states are primitive and which derived is appealing. However, most sophisticated paleontologists who basically support this view (e.g., Harper, quoted above) recommend using the fossil record only if it is relatively dense. Other paleontologists, on the other hand, are altogether sceptical: "The fossil record for most groups of organisms is too incomplete to allow the assumption that relative stratigraphic position is necessarily indicative of morphocline polarity" (Schaeffer, Hecht, and Eldredge 1972:37).

Perhaps a central problem within the traditional paleontological method has been the tendency to equate a chronocline—the sequential distribution of character-states through time—with an *ancestral-descendant sequence of taxa* possessing those character-states. Moreover, allegiance to the fossil record for determining polarity of character-states necessarily requires acceptance of the following assumptions: (a) that the record being used is sufficiently complete to make a determination of polarity (i.e., additional fossils should not materially influence the determination), and (b) that more or less axiomatically, those character-states occurring earlier are primitive. The acceptance of both assumptions seems to be a matter of faith, and perhaps the general "experience" of the investigator. But, clearly, both assumptions will always be open to doubt, and how are we to say whether they are valid for any particular case?

That there can be no prior reason against expecting primitive character-states later in the record and derived states earlier is illus-

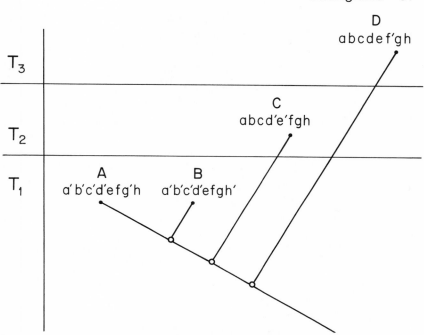

Figure 2.12 Four taxa (A–D) distributed in three different stratigraphic time intervals. Each taxon has eight characters, derived character-states being indicated by prime marks. This hypothetical example, which may not be uncommon, illustrates the danger of assuming that character-states occurring earlier in the fossil record are primitive. (See text.)

trated in figure 2.12. In this example, four species of a monophyletic group (A–D) are distributed through time (the reader is invited to supply any time scale desired) and, on the basis of synapomorphy patterns, are related as shown. The lineage A + B is characterized by the possession of a number of derived character-states (*a'*, *b'*, *c'*), whereas species C and D, which are sampled from a later time, are markedly more primitive in morphology. For some character-states (*e → e'* and *f → f'*), the fossil record would give a true picture of the polarity, but for most others that picture would be misleading.

Although this example is didactic in nature, it is nevertheless easy to see that each lineage must be characterized by one or more derived character-states, and thus *whenever sister-species have unequal survivorship in the fossil record, it is an expectation of the phylogenetic process that some taxa occurring earlier in the fossil record will possess derived character-states relative to those occur-*

ring later in the record.[4] Because unequal survivorship of sister-species is a common phenomenon, the early occurrence of derived character-states should also be common. It is also easy to see that the more incomplete the fossil record, the less confident we can be in using that record for determining polarity.

Does this argue against using paleontological data in the analysis of synapomorphy? No, not entirely. Rather, it is an argument against the axiomatic acceptance of the traditional paleontological approach. If other methods of analysis, for example outgroup comparison, are unable to provide a sufficiently clear indication of polarity for one or more characters, then in such cases, paleontological data might be used, with caution, to hypothesize a primitive-derived sequence (see also Delson 1977). This hypothesis itself will be subject to testing when alternative cladograms are compared and evaluated (see below). Even so, in most studies involving fossil taxa, outgroup comparison will yield a more reliable estimation of polarity sequences and is, therefore, to be preferred over a strict adherence to the occurrence of character-states in stratigraphic sequences. If both outgroup comparison and stratigraphic occurrence indicate the same hypothesis of synapomorphy, one's confidence in the analysis is understandably increased.

2. The application of ontogeny Early in the chapter, we applied ontogenetic data to the problem of identifying evolutionary novelties (synapomorphies) within a group of five vertebrate taxa. The discussion was presented in terms of deciphering levels of character distribution (synapomorphy) by examining the distribution of different character-states both in the adult and in early developmental stages. Thus cartilage was considered more widespread (primitive) relative to bone because the former is common to all five taxa in early stages of development and is transformed to (replaced by) bone in later ontogenetic stages in the perch, lizard, cat, and mouse. In the discussion that follows, we expand on the use of ontogenetic data in character analysis and discuss its strengths and possible limitations.

---

4. This statement is true if a paleontologist is restricting the analysis to a given monophyletic group. If outgroup comparison is used as a method to determine polarity, then at some point in the comparison it may be possible to show that the primitive character-state in fact occurred earlier in time. But for any given collection of taxa, and using the paleontological approach, we expect some derived character-states to occur in earlier strata (see text).

Historically ontogenetic observations have been considered useful for determining the course of phylogeny in two ways. First, there is the straightforward Haeckelian viewpoint that "ontogeny recapitulates phylogeny," a statement known as the biogenetic law. In this view, the ontogenetic stages of a descendant are said to trace the sequence of adult ancestors (the reader is referred to Gould 1977a, for a historical review and detailed characterization). Thus, by "reading" ontogeny, it is said, phylogeny can be reconstructed. This strict interpretation of recapitulation has long since been rejected by modern biology, yet the notion that each individual somehow has its phylogeny "locked up" in its development has continued to intrigue systematists.

The second way in which ontogenetic data have been applied to phylogenetic reconstruction is actually much older than the Haeckelian formulation. In the late 1820s the German comparative embryologist K. von Baer generalized the results of his detailed studies in a set of rules or laws.[5] These have been expressed by deBeer (1948:3) as follows:

1. In development from the egg the general characters appear before the special characters.
2. From the more general characters the less general and finally the special characters are developed.
3. During its development an animal departs more and more from the form of other animals.
4. The young stages in the development of an animal are not like the adult stages of other animals lower down on the scale, but are like the young stages of those animals.

It is doubtful that anyone since von Baer has made as important a contribution conceptually to the question of the parallel between ontogeny and phylogeny—"the most important words in the history of embryology" (Gould 1977a:56). Indeed, it is surprising that so little attention has been paid to ontogenetic data in twentieth-century systematic writings. Von Baer's fourth rule is a direct contradiction of Haeckel's biogenetic law. The first three rules pertain to the problem of analyzing synapomorphy, and our subsequent discussion focuses on them.

5. We do not mean to imply, of course, that von Baer was an evolutionist or that he interpreted his laws within an evolutionary context. The parallel between orderly ontogeny and phylogenetic inference was an interpretation of these laws initiated after 1859.

The question we shall examine here is twofold: how can ontogenetic data be used to determine the level of distribution of character-states of adult taxa? And how do we interpret ontogenetic character transformations in terms of their level of synapomorphy? Consider a monophyletic group of four species, A through D, in which A and B have an endoskeleton composed of cartilage and C and D an endoskeleton composed of bone. Which condition of the adult endoskeleton is primitive, cartilage or bone? That is, which is more widely distributed? Examination of the adult stage of these four species does not provide an answer. Assume that the ontogenies of the four species are examined and all are found to have a cartilaginous endoskeleton in the early stages of development; in species A and B that condition is retained in the adult, whereas in species C and D the cartilaginous skeleton is transformed into bone. Based upon a comparison of just these four species, we would conclude that a cartilaginous skeleton is a more general, more widely distributed feature (it is possessed by all four species). A bony skeleton is interpreted as a more restricted similarity, a synapomorphy. These conclusions support the hypothesis that species C and D comprise a subgroup within the larger group, A–D. Nothing further can be said about the existence of other subgroups—(C + D) + A, (C + D) + B, (C + D) + (A + B)—because the cartilaginous endoskeleton of A and B is interpretable as a symplesiomorphy.

Viewed in this sense, then, ontogenetic data can be examined directly for information about the relative distributions of adult character-states. A procedure such as this seems equivalent to an application of von Baer's first two rules and also of the biogenetic law as restated by Nelson (1978:327): "Given an ontogenetic character transformation, from a character observed to be more general to a character observed to be less general, the more general character is primitive and the less general derived."

The question arises as to how we are to interpret the level of character transformations themselves, and, in particular, how we are to analyze the problem of neoteny (Nelson 1978).[6] It is easily appreciated that neoteny invalidates the application of von Baer's laws and the biogenetic law (figure 2.13). In figure 2.13a, the ancestor

6. The term *neoteny*, or, more completely, "phylogenetic neoteny," as used here, is equivalent to Gould's (1977a:227–28) term paedomorphosis: features that appear only in the juveniles of ancestors later appear as juvenile and adult characters in descendants.

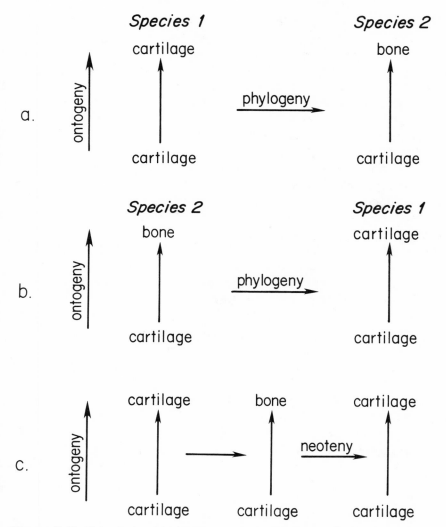

Figure 2.13 The relationship between ontogeny and phylogeny. (a) The descendant species 2 has modified its ontogeny in terms of change from a general character (cartilage) to a specific character (bone); this transformation can be intepreted with reference to the biogenetic law. (b) The descendant species 1 has modified its ontogeny in terms of change from the specific to the general; this transformation represents a case of neoteny and negates the use of the biogenetic law. (c) Neoteny represents a character reversal and thus confuses analysis of character phylogeny for that character. (See text.)

(species 1) has a cartilaginous skeleton in both juvenile and adult stages; the descendant (species 2) has evolved a transformation to a bony skeleton in the adult. The developmental histories of both species, interpreted within the framework of von Baer's laws, would lead to the phylogenetic inference that the condition in species 2 is derived relative to the condition in species 1. The occurrence of neoteny, however, would falsify the use of von Baer's laws (figure 2.13b): ontogeny cannot be interpreted as going from the general to the specific.

Suppose, in some third species descended from species 2, that the ontogenetic transformation of cartilage to bone is lost—cartilage is retained throughout the developmental sequence and becomes *secondarily* an adult character (figure 2.13c). Phylogenetically, then, a cartilage-cartilage transition is primitive for the group containing species 1 + 2 + 3, but is *also* advanced (an autapomorphy) for species 3. Neoteny produces character reversal, which creates the confusing effect of the "same" character with two different distributions (species 1 + 2 + 3, and species 3 only).

Thus for any comparative sequence of ontogenetic transformations, that sequence thought to be more general, or primitive along the usual line of argument for ontogenetic data may, in any particular example, reflect instead an instance of neoteny. Neoteny can only be detected if the analysis of other features suggests acceptance of a cladogram with an arrangement of taxa such that an adult character in fact has two separate distributions. A corroborated hypothesis of two separate distributions of what was initially taken as a single character establishes the presence of neoteny and further implies that the "character" cannot be homologous between the two distributions: retention of cartilage in the adult as a derived character (i.e., with bone suppressed) is not the same, in the evolutionary sense, as simple retention of the primitive state. Since the characters cannot be told apart on inspection, only demonstration of conflicts of distribution can be used to detect their presence as two discrete characters. Thus neoteny is a convergence (or parallelism) and is resolved in precisely the same fashion as in all such instances: two characters taken to be the "same" (homologous) can be shown to be convergent only if the analysis of further characters demonstrates that the taxa sharing the resemblance are in fact more closely related to some other taxa which do not share the resemblance (see page 70 for further discussion of the analysis of parallelism and convergence).

3. The application of outgroup comparison The most commonly applied method of determining the relative distributions of two or more character states (or determining their "polarity") is the procedure of outgroup comparison. It is based upon a rather simple methodological principle: if we are attempting to resolve the scheme of synapomorphy within competing three-taxon statement cladograms, those character states occurring in other taxa within a larger hypothesized monophyletic group *that includes the three-taxon statement as a subset* can be hypothesized to be primitive (plesiomorphous) and those character-states restricted to the three-taxon statement itself can be hypothesized to be derived (apomorphous). In short, the procedure is one of mapping the distributions of characters-states within and without the group under consideration. This principle of outgroup comparison can be illustrated with a simple example. In figure 2.14a and b are two alternative three-taxon statements. In figure 2.14a taxa A and B share a similar character-state *a'*, whereas in figure 2.14b taxa A and C share a character-state *b'*. If taxa A, B, and C are themselves members of a still larger monophyletic group A–E, then outgroups, that is, taxa outside the three-taxon statement, can be compared for the possible occurrence of alternative character-states, in this case *a* or *a'* and *b* or *b'*. Those character-states found in the outgroups, then, are considered *primitive within the three-taxon statement.* If, for example, taxa D and E possessed character-states *a* and *b'* (see figure 2.14c), this would suggest the hypothesis that character-states *a'* and *b* are derived, and this in turn would lead us to prefer the cladogram of figure 2.14a, because the postulated synapomorphies are consistent with its branching pattern. In an evolutionary context, we would postulate the evolution of *a → a'* between ancestors Y and X, and *b' → b* within the lineage leading to taxon B.

This procedure, then, can be viewed as a search for levels of character distribution in the observed similarities. In this sense, similarity *a'* would have a distribution (define a set) at the level of taxon A + B, and *b* at a still lower level, taxon B. Character-states *a* and *b'*,

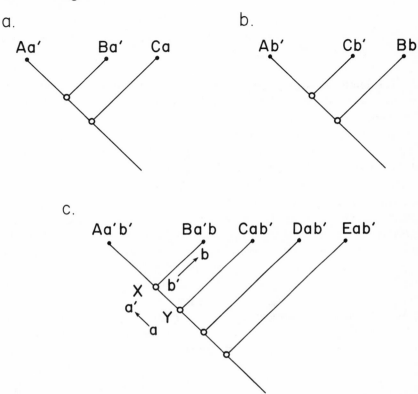

Figure 2.14 An example of outgroup comparison. (a and b) Two apparently incompatible character distributions based on shared similarities (*a'*, *b'*). (c) Comparison of outgroups D and E suggests that *a'* is a synapomorphy and *b'* a symplesiomorphy. (See text.)

on the other hand, would have a distribution *at least* at the level of taxon A–E and perhaps higher (this would be subject to independent investigation).

The use of outgroup comparison is a search for the hierarchical level of distribution of each character-state. Put another way, the method is a search for the level of the hierarchy at which each character-state is a synapomorphy and therefore defines a set of taxa. At all lower levels that character-state is a symplesiomorphy.

So how were levels of similarity, and thus synapomorphies, determined for the character-states of the present example? Why were D and E chosen for the outgroups in contrast to the potentially large number of taxa that might have been chosen? Examination of the three-taxon statement shows that character-state *a'* defines the group A + B, and a character-state *b'* the group A + C. Clearly, to resolve the conflict in the level of synapomorphy of these two similarities, additional taxa must be examined. If more As, Bs, or Cs were added to the analysis, the distribution of similarities would probably remain pretty much the same and we would consequently be dealing instead only with basic taxa of higher taxonomic rank than was done initially. It is necessary to investigate taxa that do not belong to these three groups, because the strategy is to discover whether *a* or *a'* and *b* or *b'* are found in other taxa at a higher hierarchical level. But how was the level A + B + C determined in the first place? What are their possible synapomorphies? Thus, if A, B, and C were the cat, mouse, and lizard, using our example at the beginning of the chapter, we might make a preliminary definition of this hierarchical level by saying they are vertebrates with an amniote egg. This immediately leads us to ask what are the distributions of character-states *a, a', b,* and *b'* in vertebrates lacking an amniote egg. We might observe, for example, that frogs and fishes exhibit character-states *a* and *b'* and thus we conclude that *a'* and *b* are derived within amniotes.

To summarize: in order to determine the level of synapomorphy for a character-state within an initial three-taxon statement, it is necessary to compare taxa at a higher level. This is accomplished, ultimately, by using other similarities to erect a preliminary hypothesis about the group-membership of that higher level. It would seem that it is not possible to decipher the level of synapomorphy of one similarity without tentative acceptance of a higher level of synapomorphy defined by a different similarity. In systematic work of this kind, one is always accepting, as working hypotheses, higher level hypotheses of synapomorphy (and thus relationship) as a means of evaluating lower level hypotheses. All levels of hypotheses are continually subject to testing and reevaluation (see below).

Within the systematic literature, outgroups are traditionally described as being the "closest relatives" of the taxa within the three-taxon statement. In many systematic studies, a hypothesis about those "close relatives" frequently has been corroborated, or at least the range of possibilities has been narrowed to a small number of taxa. Choosing outgroups to be compared is usually not as difficult a task as some critics of cladistic analysis would have us believe. For, after all, it is rare that a given three-taxon statement cannot be considered a subset of some higher level hypothesis of relationship, and therefore there will always be some basis for evaluating hypotheses of synapomorphy. Contrary to these criticisms, it is not necessary to specify the close-relative exactly but only a higher level of synapomorphy. This does not mean, of course, that the determination of synapomorphy will be easy, or even possible in some cases. But this is not a limitation of outgroup comparison, or of cladistic analysis in general. Rather, it is a limitation set by the available data, a limitation of our knowledge, as it were, and such a situation would seemingly jeopardize all methods of systematic analysis equally. Some examples and problems of outgroup comparison will be presented later in the chapter.

Character weighting It is our view that each hypothesis of synapomorphy, if arrived at by careful comparative analysis, can contribute to the evaluation of alternative cladograms. Recently, however, the opinion has been expressed that some hypotheses of synapomorphy are less useful and should have less weight than other hypotheses (Hecht 1976; Hecht and Edwards 1976, 1977). This viewpoint has also gained some approval from evolutionary systematists (Szalay 1977). It is thoroughly consistent with the prevalent tradition in evolutionary systematics to set up a system of weighting whereby some sorts of characters are deemed axiomatically to be more (or less) useful in phylogenetic analysis than others. We merely reiterate the fundamental observation that all characters are evolutionary novelties (synapomorphies) at some level. The problem is to find that level for each character, and not to assume that "conservative" features are more or less valuable than "variable" ones, or that "non-adaptive" features are better or worse than obviously "adaptive" features, just to name two different sorts of characters occasionally deemed important as criteria for the purposes of phylogenetic inference. It is still commonly said, for instance, that obviously functional ("adaptive") features are more likely to represent convergences or parallelisms than are structures whose functions are less obvious. But parallelisms and convergences can only be hypothesized by arguing that one or more of the taxa with that structure are more closely related to some organism without that structure. Because the function of a structure is assumed to be "understood" is no reason to assume that it evolved more than once. Such an assumption reflects the preoccupation with adaptation predominant in contemporary evolutionary theory (chapter 6). We return to the analysis of parallelism and convergence later in this chapter.

Synapomorphy as a Test of Cladistic Hypotheses

In previous sections of this chapter, we presented the notion that cladograms are hypotheses about synapomorphy patterns, and that, given a unique pattern to the history of life, there is an expectation that a cladistic hypothesis which closely approximates, if not precisely parallels, the unique historical pattern should exhibit maximal congruence of the nested synapomorphies. If one views the science of systematics as being subject to the same rules of inference as other branches of hypothetico-deductive science, then these assumptions and expectations take on the nature of an axiomatic methodological principle: because we cannot empirically have knowledge of the true historical pattern, science must formulate a criterion by which to judge the relative merits of our close approximations (hypotheses). That criterion, in effect, is parsimony, and it specifies the most preferred hypothesis to be the one exhibiting the most congruence in the synapomorphy pattern. This is not to suggest that this hypothesis is necessarily true or that it precisely mirrors the true historical pattern, but only that it appears to be a better hypothesis than the alternatives. As Wiley (1975:236) so aptly remarks, the application of parsimony must be accepted, "not because nature is parsimonious, but because only parsimonious hypotheses can be defended by the investigator without resorting to authoritarianism or apriorism."

Synapomorphies are tests of cladistic hypotheses (cladograms), but reciprocally these cladograms, with their expected pattern of congruent synapomorphies, are also tests of the synapomorphies. Such reciprocity is an integral part of hypothetico-deductive science, particularly if one views all scientific statements—even those considered to be "facts"—as being theory-laden (see Popper 1959, and many other philosophers of science). Detailed discussions about the application of synapomorphy to test cladistic hypotheses can be found in Wiley (1975), Gaffney (1979), and papers cited therein. We will describe some basic principles of this procedure.

Consider the three alternative (competing) three-taxon state-

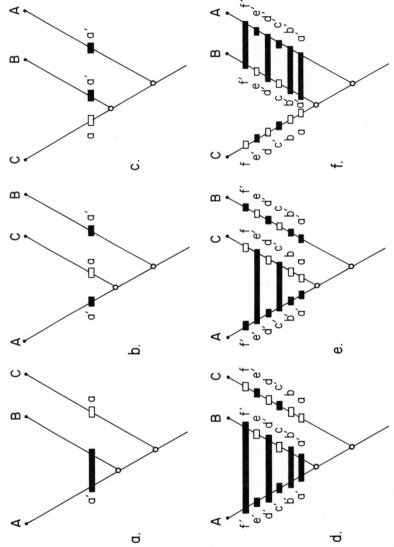

Figure 2.15 Testing alternative cladograms with postulated synapomorphies. Shaded rectangles indicate synapomorphies; open rectangles indicate the primitive condition. (See text for details.)

ments in figure 2.15. If there is a postulated synapomorphy, *a'*, present in taxa A and B, whereas the primitive condition, *a*, is in C, then we can say that cladogram 2.15a is corroborated and cladograms 2.15b and 2.15c are rejected.[7] A decision about corroboration or rejection is dependent upon the ability of the synapomorphy to define a set within each cladogram.

If similarity *a'* is truly a synapomorphy within the three-taxon statement A + B + C, then the foregoing argument is clearly acceptable. But, what if *a'* is not a true synapomorphy within A + B + C? How do we account for this observed similarity in A and B? There seem to be three, and only three, possibilities and the first one really does not count: (1) The investigator has not actually perceived a similarity *a'* at all; a mistake has been made. This is the least likely of the possibilities, for after all, if *a'* can be recognized and named, then some similarity must be apparent. This leaves us with two alternatives: (2) *a'* is in fact a true synapomorphy but it defines a set at a higher level of the hierarchy, say A + B + C, or even higher. If this is the case, then the shared similarity *a'* in A and B is a symplesiomorphy and, *a* in taxon C is almost certainly to be interpreted as an autapomorphy. (3) The similarity *a'* in A and B is not a synapomorphy at any level of the hierarchy, therefore it must be interpreted as a convergence, in other words the similarity was independently acquired during the evolution of A and B. If this is true, the condition *a* in C cannot be analyzed given present information. It should be clear at this point that a convergence is the antithesis of homology; it is neither a synapomorphy nor a symplesiomorphy.

The preceding discussion inevitably leads to the question: if we have perceived a similarity distributed in two taxa of a three-taxon statement, how do we decide whether a postulated synapomorphy is a "true" (corroborated) synapomorphy or a convergence? The question can be restated: how do we corroborate or reject a hypothesis of synapomorphy?

In order to answer this question, let us return to our example of figure 2.15. If a second postulated synapomorphy also defines the

7. In the systematic literature, one will also find the word "falsified" to refer to hypotheses such as those represented by cladograms b and c. "Falsified" implies that the hypotheses are proven false, but this is not the meaning we (or other phylogenetic systematists) wish to convey. It may be that, eventually, the preferred hypothesis will itself be "rejected" by some synapomorphies. Another word that is occasionally used is "refute," but this has essentially the same implications as "falsify."

subset A + B, the cladogram of a is again corroborated whereas b and c are rejected. It is clear, then, that continued testing of the cladograms will tend to increase the degree of corroboration or rejection of each hypothesis. Let us assume, for example, that after repeated testing we arrive at a situation similar to that shown in figure 2.15d–f. There are six postulated synapomorphies (a'–f'), of which four (a', b', d', and f') define the set A + B (figure 2.15d) and two (c', e') define the set A + C (figure 2.15e). No postulated synapomorphies appear to define the set C + B (figure 2.15f). On this basis, then, an investigator would conclude that hypothesis f is the most rejected hypothesis, e the next most rejected hypothesis, and d the least rejected hypothesis. The criterion of parsimony specifies our acceptance of the least rejected hypothesis. If our comparative analysis of outgroups has clearly indicated that the postulated synapomorphies a'–f' could not be interpreted as being synapomorphies at a higher hierarchical level, and thus primitive within A + B + C, then

1. if the cladistic hypothesis of 2.15f were true, it would necessitate six cases of convergence (a'–f');
2. if the cladistic hypothesis of 2.15e were true, it would necessitate four cases of convergence (a', b', d', and f'); or
3. if the cladistic hypothesis of 2.15d were true, it would necessitate two cases of convergence (c' and e').

In evolutionary terms, therefore, we prefer the cladistic hypothesis which minimizes convergence. Once a preferred cladistic hypothesis has been chosen, it in turn serves as a basis for evaluating (testing) the synapomorphies. If, for example, the hypothesis 2.15d is preferred, then postulated synapomorphies a', b', d', and f' are accepted as synapomorphies at the given hierarchical level.

Such seems to be the role of synapomorphies in testing cladograms, and cladograms in testing synapomorphies. The discussion also leads us to a more detailed consideration of conflicts in synapomorphy pattern.

Conflicts in Synapomorphy Pattern and Their Analysis

The preceding section strongly implies that the identification of a shared character-state as a convergence is strictly a matter of whether the similarity can be used to define a set within a specified cladogram. Indeed, the traditionally accepted definition of convergence refers to similarities independently acquired and not derived from a common ancestor. By definition, then, one must have a hypothesis of relationships prior to applying the concept of convergence. Thus, in figure 2.15, character-state a' was recognized as a homology (synapomorphy) because cladogram d was eventually accepted. If either cladograms e or f had been accepted, then a' would have been considered a convergence. As we said, this concept of convergence has been standard in the systematic literature, although phylogenetic systematists have only recently made its formulation more explicit.

There is a belief among some systematists that it is somehow possible to assess the probability that certain similarities are the result of convergence or parallelism. It is said, for example, that those similarities with a high probability of being convergent or parallel can be eliminated from the analysis, or at least given "low weight." Such is the rationale behind the weighting scheme of Hecht and Edwards (1977). We will now comment specifically on the problem of analyzing parallelism and convergence, because this bears directly on the problem of testing cladistic hypotheses.

To many systematists (e.g., Mayr 1969:226–28) convergence of features does not present a major analytical problem because study of a sufficient number of characters usually reveals that the two taxa possessing the similarity are in fact not related to each other but to some other group. We agree and further assert that the same applies to the concept of parallelism. Other systematists, on the other hand, believe that parallelism is different in important ways. What do these workers mean by parallelism? Some definitions follow.

> Parallelism is the development of similar characters separately in two or more lineages of common ancestry and on the basis of, or channeled by, characteristics of that ancestry. (Simpson 1961:78)
>
> [Parallelisms are] similarities resulting of joint possession of independently acquired phenotypic characteristics produced by a shared genotype inherited from a common ancestor (similarity through parallel evolution). (Mayr 1969:202)
>
> [In] parallel evolution . . . the character is present in the ancestral form but a common derived character state has been independently evolved in each descendant form. (Hecht and Edwards 1976:654)

The concept of parallelism is most easily compared and contrasted with convergence by reference to branching diagrams (figure

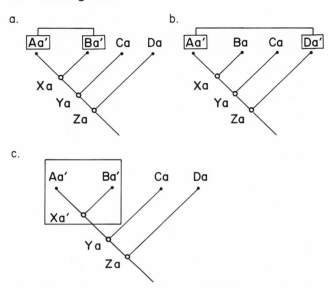

Figure 2.16 The concepts of parallelism and convergence. (a) An example of parallelism. (b) An example of convergence. (c) An alternative hypothesis to parallelism in which the ancestral condition in X is postulated to be *a'* because the sister-groups A and B both have *a'* (See text).

2.16). The condition of the similarity in the common ancestor is the central question in applying the two concepts. In parallelism (figure 2.16a), the common ancestor X is assumed to have a relatively more primitive condition of the character than its descendants, and a similar derived condition is assumed to have evolved in each descendant lineage. In convergence (figure 2.16b), on the other hand, the common ancestor Z is also assumed to have the primitive condition, but the descendant taxa having the derived condition are not sister-groups of one another. This would seem to be the only meaningful distinction between the two concepts. Hecht and Edwards (1976, 1977), for example, consider the atlas-axis complex of birds and mammals to be a parallelism, but because the two taxa are so distantly related, their example would seem to vitiate all useful distinction between parallelism and convergence.

There is, however, a series of assumptions surrounding the application of parallelism that are seldom, if ever, applied to convergence. In parallelism, it is argued, because closely related species have similar genetic backgrounds and therefore similar developmental potentials, they are likely to respond to similar selection forces in a like manner; hence parallelism is to be expected in close relatives living under similar environmental conditions. We might, for the moment, accept this line of argumentation and ask: how are we to analyze such a situation in systematic studies?

In order to apply this concept of parallelism (say, as in figure 2.16a), a minimum of four sets of information is required:

1. It must be known or assumed that the common ancestor X in fact had the primitive character-state *a*.

2. It must be known or assumed that the taxa (or individual organisms) of the lineages leading to A and B had genetic and developmental potentials which were sufficiently similar to some specified degree.

3. It must be known or assumed not only that the taxa of A and B live in similar enviornments, but that they were subjected to similar selection forces.

4. It must be known or assumed that if they were subjected to similar selection forces, these taxa would in fact respond to them in similar ways.

Given the above four assumptions, one could conclude that parallelism as defined by evolutionary systematists had occurred.[8]

It is our opinion that this conception of parallelism is epistemologically impossible to evaluate and therefore unscientific. To apply this concept—and, accordingly, a weighting scheme—to the analysis of synapomorphy necessitates the inclusion of numerous ad hoc assumptions. Thus, one cannot test any hypothesis of parallelism without resorting to knowledge about ancestral conditions, genetic and developmental potentials, the selection forces present in the past, and the expected response to those selection forces.

The concept of parallelism is also open to criticism on the grounds of parsimony. An argument of parallelism requires that two evolutionary lineages evolved a derived character-state independently. A properly expressed cladistic hypothesis, on the other hand, would necessitate the evolution of the derived condition only once,

8. In actuality, all that would be required to invoke parallelism would be knowledge of relationships and knowledge that the common ancestor X has the primitive condition. Most workers, however, especially evolutionary systematists, also include assumptions about developmental–genetic potential and natural selection.

between common ancestors Y and X (figure 2.16c); the unspecified common ancestor X would be inferred to have the derived condition *a'*.

In conclusion, we recommend that the concept of parallelism be omitted from systematic studies. We suggest that the term convergence be applied to all cases of nonhomologous character similarities, identified through character conflicts arising from normal analysis of the patterns of synapomorphy.

Internesting Statements of Monophyly

One can view the problem of cladogram construction as a problem of internesting two or more three-taxon-statement hypotheses of monophyly. Any monophyletic group can be resolved to a basic taxonomic unit of a cladogram at a higher hierarchical level (see Wiley 1975). For example, in figure 2.17 the monophyletic group A + B can be represented by the taxon-name W at a higher level of the hierarchy (figure 2.17a), W + C by X at a still higher level (figure 2.17b), and so on (see also Gaffney 1979).

All levels of the hierarchy are susceptible to testing by postulated synapomorphies. Each identified synapomorphy can test at only one level of the hierarchy. Furthermore, to test at any one level, that is, to make a determination of synapomorphy, requires comparison of taxa at higher levels. It can be appreciated, therefore, that the choice of an outgroup at any one level of comparison is itself a hypothesis of monophyly to be tested at still higher levels.

Some Case Studies

In this chapter, we have attempted to provide systematists with a methodological foundation for formulating their own cladistic hypotheses. Most of the examples were hypothetical, and while such theoretical taxa as A, B, and C have their value, real-world examples also have their own advantages. Thus, we close this chapter with some examples of cladistic analysis. The presentation of examples is a difficult task in a general book of this kind because the readership

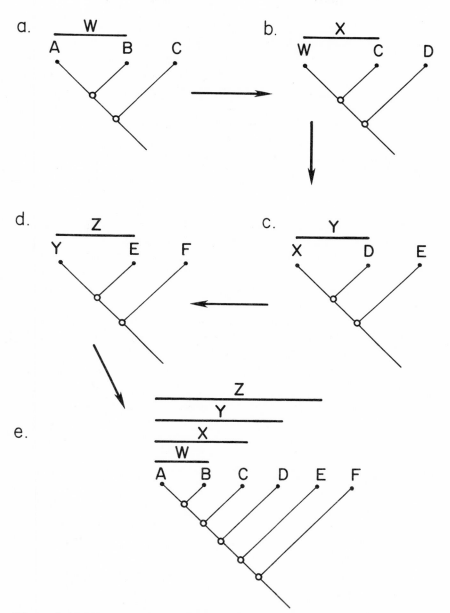

Figure 2.17 The resolution of lower-level hypotheses of monophyly to higher-level hypotheses (see text).

will have diverse interests and experience. To properly evaluate any given study requires not only familiarity with the comparative biology of the organisms themselves but also a little experience in applying the theory we have discussed up to this point. Nevertheless, we hope the examples will be of value, and that they will give a general idea of the procedures and kinds of evidence used in cladistic analyses. Most of the studies begin with a comparison of similarity, followed by an analysis of synapomorphy, and postulated nested sets of taxa are then constructed on the basis of the postulated synapomorphies.

Example 1. Actinopterygian fishes Wiley (1976) recently performed a cladistic analysis on some major groups of actinopterygian fishes. His main concern was the relationships of the species within the family of gars (Lepisosteidae) and the relationships of that family to other groups of actinopterygians. Our discussion here is restricted to the latter portion of his study.

The relationship of gars to other actinopterygians has constituted a systematic problem for many years. Wiley identified four cladistic hypotheses that have been suggested frequently in the literature (figure 2.18), and then proceeded to evaluate each using postulated synapomorphies. The similarities studied were primarily those of the cranial and postcranial osteology, but some myological similarities were noted. The determination of the level of synapomorphy for each similarity is seldom easy within the higher taxa of fishes, and in formulating his hypotheses Wiley compared a broad range of taxa, both actinopterygian and nonactinopterygian.

The results of Wiley's analysis are shown in figure 2.19. A large number of the postulated synapomorphies indicate that amiids (bowfins) can be placed in a group along with the teleosts (synapomorphies 1–13) and that this group, in turn, is the sister-group of the gars (synapomorphies 14–20). Not shown in this figure are the synapomorphies linking gars, amiids, and teleosts to the chondrosteans; the reader should consult Wiley (1976:14–37) for his detailed rationale regarding determination of synapomorphies.

Wiley's study is noteworthy because it is a cladistic analysis in which a set of specific hypotheses was first formulated on the basis of prior systematic work, and then each hypothesis was tested by the results of an analysis of synapomorphy. In this case, the synapomor-

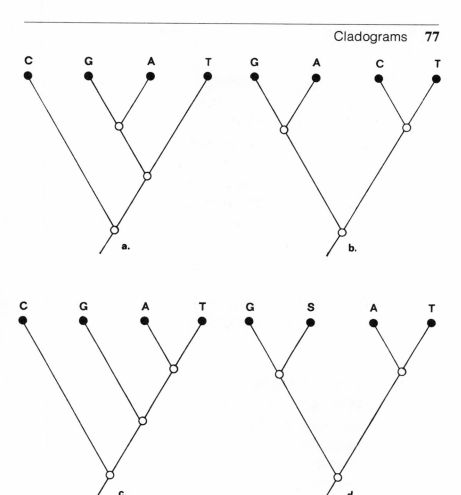

Figure 2.18 The four cladograms subjected to testing by Wiley in evaluating the interrelationships of some major groups of actinopterygian fishes. A: amiids; C: chondrosteans; G: gars; S: semionotids; T: teleosts. (From Wiley 1976:14.)

phies themselves were not used directly to construct internesting statements of monophyly.

Example 2. Side-necked turtles Gaffney (1977) has analyzed the synapomorphy patterns of cranial similarities for seven genera of chelid turtles. The family itself is well-defined by synapomorphies within turtles as a whole, thus the study was initially restricted only to this family. Gaffney used all other turtles as a basis for evaluating synapomorphies within the chelids; in addition, this study was interpreted

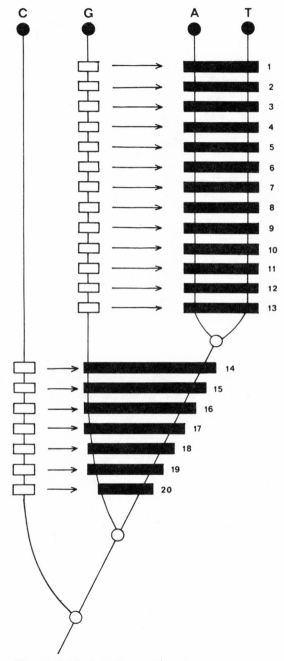

Figure 2.19 The preferred cladogram of actinopterygian fish interrelationships. Synapomorphies are indicated by shaded rectangles. A: amiids; C: chondrosteans; G: gars; T: teleosts. (From Wiley 1976:38.)

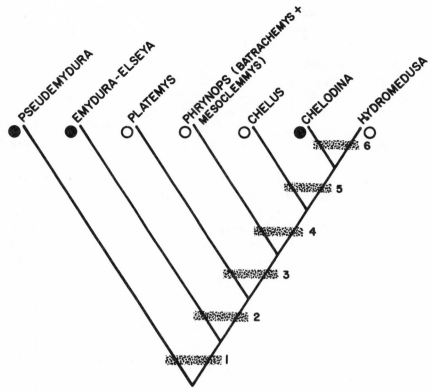

Figure 2.20 A cladistic hypothesis for the genera of turtles in the family Chelidae. (From Gaffney 1977:17, figure 10.)

within the framework of a prior cladistic analysis of all major groups of turtles (Gaffney 1975). For example, *Chelodina* and *Hydromedusa* are the only two genera of chelids to have four claws on their forefeet (all other genera have five). Moreover, the sister-group of the Chelidae, the Pelomedusidae, also have five claws. Therefore, the condition in *Chelodina* and *Hydromedusa* was postulated to be derived.

Gaffney's results are summarized in figure 2.20. Six sets of synapomorphies were found to define six sets of nested taxa. In this example, because of a paucity of previous work on chelids, well-defined alternative hypotheses were not recognized prior to the analysis, as was done in the preceding example. Thus Gaffney proposed hypotheses of synapomorphy and then used these to internest various three-taxon hypotheses of monophyly.

Example 3. Ratite birds The large, flightless ratite birds have excited considerable comparative work for over a century, but almost no studies postulated interrelationships because, until very recently, few workers considered them to be a monophyletic group. Following prior work arguing for their monophyly (Bock 1963; Parkes and Clark 1966), Cracraft (1974a) constructed a cladogram of nested synapomorphies. This study concentrated on cranial and postcranial similarities, but the results were congruent with synapomorphies postulated for similarities in behavior and breeding biology (Meise 1963).

Because one or more ratite taxa have frequently been considered the possible sister-group of all other birds, their study presented some special problems for determining the level of synapomorphy of the different character-states. If ratites are the sister-group of all other birds, this would necessitate extending the comparison beyond birds to other vertebrates, particularly diapsid reptiles. But the attributes of these nonavian groups are not comparable to the similarities of ratites and other birds in sufficient detail to make such a comparison useful (the detailed similarities of ratites and other birds do not, generally speaking, extend to diapsids). Hence, it was decided that those similarities within ratites having widespread distribution in all other birds, particularly nonpasseriforms, would be hypothesized to be synapomorphous at a hierarchical level higher than ratites (i.e., to be symplesiomorphous within ratites). Such a procedure seemed to work, for it was discovered that seven sets of postulated synapomorphies defined nested subsets for the eight included taxa (figure 2.21), and no other cladogram was suggested by the data. As an example of the reasoning used to postulate synapomorphies, we can examine the morphological trends seen in the tibiotarsus, one of the hindlimb bones (figure 2.22). In primitive ratites (e.g., the tinamou *Crypturellus*), and in nonpasseriform birds in general, the cnemial crests of the tibiotarsus are bladelike structures and not greatly enlarged. Not only is there a trend within ratites for the crests to become enlarged anteriorly, but their base becomes strongly constricted (figure 2.22F) and the outer cnemial crest becomes knoblike. This latter feature is unique, and clearly derived, and was interpreted as a synapomorphy of *Rhea* and *Struthio*. Other synapomorphies were postulated in a similar manner.

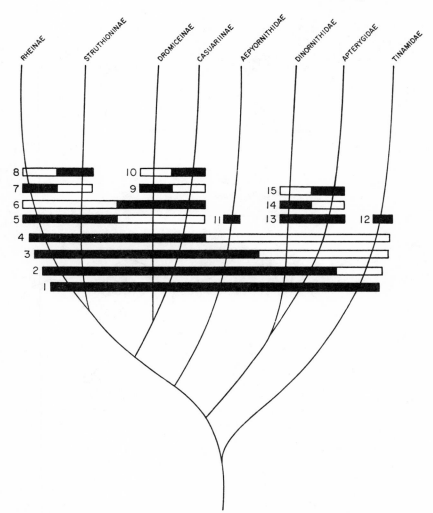

Figure 2.21 A cladistic hypothesis for the families of ratite birds.
(From Cracraft 1974:495, by permission of the British Ornithologists'
Union.)

Example 4. Hominoid primates In recent years cladistic analysis has
been applied to the question of primate interrelationships, and our
example here pertains mostly to some studies of the great apes and
man. In particular, the reader can consult Delson and Andrews

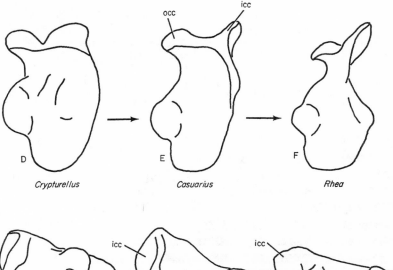

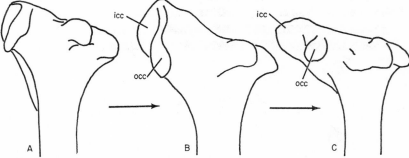

Figure 2.22 Hypothesized primitive–derived sequences in some similarities of the proximal tibiotarsus of ratite birds. The condition of the tinamou *Crypturellus* is similar to that of other nonpasseriform birds and is thus postulated to be primitive (see text). (From Cracraft 1974:501, figure 7, by permission of the British Ornithologists' Union.)

(1975), Eldredge and Tattersall (1975), Delson (1977), and Delson, Eldredge, and Tattersall (1977). All of these papers not only include a discussion of the taxa and the synapomorphies used to unite them, but each also presents useful summaries and statements about the theory and methodology of cladistic analysis. These papers are also good examples of analyses combining data from both fossil and recent taxa; in hypothesizing polarity of character-states the authors used outgroup comparison (to other, "lower" primates) and stratigraphic distribution. A representative cladogram of the higher primates is shown in figure 2.23 (after Delson 1977).

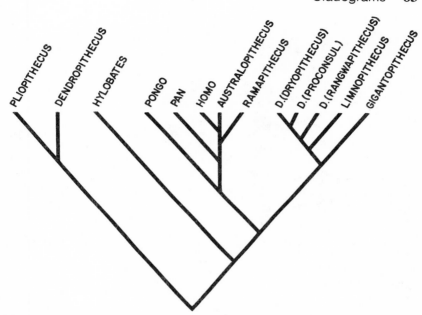

Figure 2.23 A cladistic hypothesis for the great apes and man. [With permission from Delson 1977:445, figure 1. Copyright by Academic Press Inc. (London) Ltd.]

Example 5. Spiders Platnick and Gertsch (1976) undertook a cladistic analysis of the major groups of spiders, and their paper dealt almost exclusively with the three-taxon statement including the taxa Liphistiidae, Mygalomorphae, and Araneomorphae. In order to determine which similarities are primitive and which derived, they presented evidence for the arachnid order Amblypygi being the sister-group of the spiders as a whole. They then hypothesized "that any character-state found in some but not all spiders and also in amblypygids is plesiomorphic, and its homologs apomorphic" (p. 2). In addition to outgroup comparison, Platnick and Gertsch employed ontogenetic data, effectively following von Baer's principles to define the polarity. As an example of their approach to character analysis, they noted that liphistiid spiders have the third abdominal segment as a distinct sclerite ventrally. The Mygalomorphae and Araneomorphae, on the other hand, have lost all ventral indication of segmentation. Because the amblypygids have a distinct ventral sclerite, the condi-

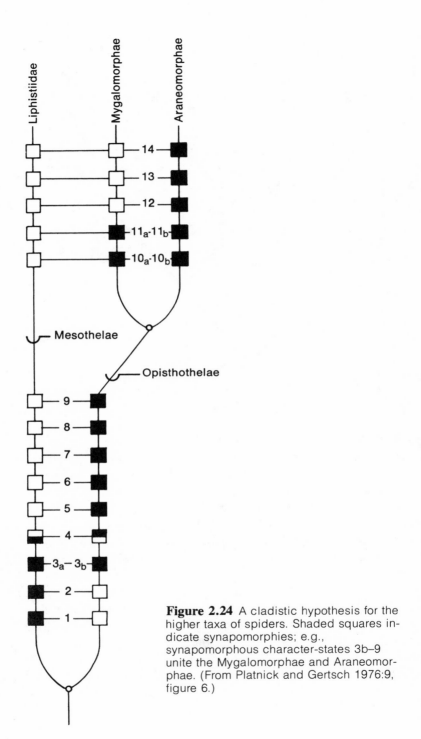

Figure 2.24 A cladistic hypothesis for the higher taxa of spiders. Shaded squares indicate synapomorphies; e.g., synapomorphous character-states 3b–9 unite the Mygalomorphae and Araneomorphae. (From Platnick and Gertsch 1976:9, figure 6.)

tion in mygalomorphs and araneomorphs was postulated to be derived. The results of a similar analysis on 14 characters are shown in figure 2.24. For another application of cladistic analysis to the study of spider relationships, the reader is also referred to Platnick (1977c).

Chapter
3

Species: Their Nature and Recognition

THE ENGLISH word "species" derives from the identical Latin word "species." In Latin, the word means "kind." The primary purpose of this chapter is to clarify the meaning of the English word as used in modern comparative biology.

Species Defined

Linnaeus (1758) applied the term "species" in an explicitly Latin sense: species are particular kinds of generic entities. In other words, "genera" were more general entities and species were more specific kinds of those entities. This original biological usage remains a part of our general concept of the nature of species today.

The notion of organic evolution, defined for the moment as "descent with modification," leads directly to an expectation of nested sets of evolutionary novelties. These are the synapomorphies discussed in the previous chapter. Nested sets of synapomorphous resemblances are in turn used to define and recognize nested sets of monophyletic groups, or taxa. The basic aim of cladistic analysis is to define nested sets of monophyletic taxa on the basis of nested sets of evolutionary novelties. Consonant with Linnaeus' original use of the term "species," we would expect there to be a base level, where patterns of synapomorphous resemblance linked individual specimens into clusters which are not further subdivisible. These indivisible clusters would form the base of the hierarchy of nested taxa. In modern garb, they would conform exactly to the Linnaean concept of species as the basal units of the taxonomic hierarchy.

This essentially analytic view of the nature of species conforms to the notion of organic evolution as descent with modification. There is no problem with the concept as long as evolution is regarded strictly as the transformation of characters, and taxa are viewed as hierarchically arranged clusters of individuals, the hierarchy being determined by the apparent pattern of nested sets of evolutionary novelties. Two simple definitions of species consistent with this concept follow: "species are minimal monophyletic groups," or "species are taxa of the lowest categorical rank within the Linnaean hierarchy."

There is a distinct advantage to such definitions of species. As merely the lowest-ranked category of the Linnaean hierarchy, taxa of this rank are recognized in precisely the same fashion as are more inclusive (higher-ranked) taxa. The general principles adduced in chapter 2 would suffice, and species recognition would pose no special problem. Were the situation that simple, we could proceed immediately to some practical examples of species recognition and associated taxonomic procedures. But the situation is not that simple.

Factors complicating the very notion of what species are arise from several considerations. There has long existed the notion that species are somehow "real," i.e., actual "entities," "things," or "individuals" actually existing in nature. Their status as "entities" is different from their status as "kinds." (See Ghiselin 1974, and Hull 1976 and 1978 for the development of this theme from a philosophical point of view.)

The notion that species are "real" units in nature is an old one. From ancient times, and as an ongoing cross-cultural experience today, it is noted that, within a local area, there seem to be a number of different kinds of organisms. These kinds are clusters of individuals all more or less similar in appearance and behavior. These separate kinds are all fairly easily identified (i.e., they are distinct). The individuals comprising each distinct kind exhibit some pattern of reproduction among themselves, producing still more individuals of like kind. In other words, in any one area, there are different kinds of organisms organized into discrete reproductive communities. Moreover, each of these units may be found in adjacent areas, where they are either identical or slightly different in appearance. In any case, the common experience of mankind, including our perception of ourselves, is that there are discrete reproductive units in nature. Nearly all contemporary definitions of species stress their status as discrete reproductive entities. Species, moreover, are expected to have a finite distribution in space and through time.

Returning to the view that species are those taxa of the lowest rank in the Linnaean hierarchy, or, in a cladistic sense, minimal monophyletic groups, the resultant picture can be neatly diagrammed in a cladogram. But in a cladogram all taxa are terminal, whereas the general concept of evolution implies a pattern of ancestry and descent. What kinds of units are ancestors and descendants? A purely transformational view of evolution as "descent with modification" does not specify what the units are. The literature of evolutionary biology is not very helpful on this point, as there is no clear consensus: some authors think that codons are the units of evolution, and there are all shades of opinion ranging from these genetic loci to structures, individuals, populations, species, and on up the Linnaean hierarchy. It is not uncommon, for example, to hear of one phylum being ancestral to another. The Procaryota are frequently alleged to be the ancestor of the Eucaryota. It would appear that a clarification of the nature of the units of evolution would enable an improvement in existing evolutionary theory; we pursue this theme in detail in chapter 6. But the identification of the units of evolution—i.e., what kinds of units are the ancestors and descendants implicit in the very concept of organic evolution?—also bears directly on the nature of species.

Consideration of the ecological reality of populations and species and the nature of the reproductive plexus binding these units together, implies a functional hierarchical organization of molecules, codons, cells, tissues, organs, individuals, populations, and species. A codon can change (i.e., undergo mutation) within an individual. Somatic mutations rest with the individual; mutations in germ cells may be transmitted to other individuals reproductively. This is true for all genetic changes and their phenotypic manifestations. Such changes can have import in an evolutionary context only to the extent that they are, in fact, inherited. The conclusion is obvious: Genetic and phenotypic heritable change assumes evolutionary importance (persistence in time) only to the extent that the reproductive community and its descendants persist. For organisms among which there is at least occasional sexual reproduction, this unit would conform to the reproductive concept of species. We are led to the ineluctable conclusion that species, when conceived of as reproductive units, are the units of evolution.

Returning to the notion of phylogenetic ancestry and descent, if species are evolutionary units, it follows that some must be ancestral to others. In other words, not all species can be viewed as terminal taxa. In terms of the distribution of characters, a descendant unit must by definition possess at least one derived attribute not shared with its ancestor. Logically a species cannot have any synapomorphies unique to itself (i.e., autapomorphies) if it is ancestral to any other species. To the extent that each species in a collection of organisms appears to possess one or more autapomorphies, we must judge that either (a) all species sampled are terminal taxa or (b) their descendants are not included in the sample.

We conclude from the foregoing that strict application of cladistic methodology to the problem of species recognition is not entirely consistent with the concept of species as individuals and evolutionary units. Use of patterns of synapomorphy to delineate monophyletic groups requires that all taxa be terminal, whereas the very notion of evolution requires some unit—and it must be the species—serve as ancestors and descendants. Strict adherence to cladistic methodology may tend to underestimate the true number of species sampled. A corollary, and somewhat ironic, observation is that species are not always monophyletic units in a cladistic sense.

Thus far we have developed a composite view of the nature of species which includes at least two dual aspects. (1) Species are simultaneously the lowest ranked taxa of the Linnaean hierarchy (i.e., they are placed in the "species" category) *and* real entities in nature, representing the highest rung of the hierarchy of molecules, through individuals and populations. As we have developed elsewhere in this book (chapter 6), taxa of higher rank do not share with species the quality of being individuals. Rather, taxa of higher categorical rank are collections of species (see note 2 to chapter 2 for an important exception). In our view, it is important that they be monophyletic. (2) A second duality sees species as integrated reproductive communities—i.e., they exhibit a pattern of parental ancestry and descent—which gives them internal cohesion and separates them from other such groups. Simultaneously, species are evolutionary units exhibiting a pattern of phylogenetic ancestry and descent—i.e., they give rise to units of like kind (other reproductive communities).

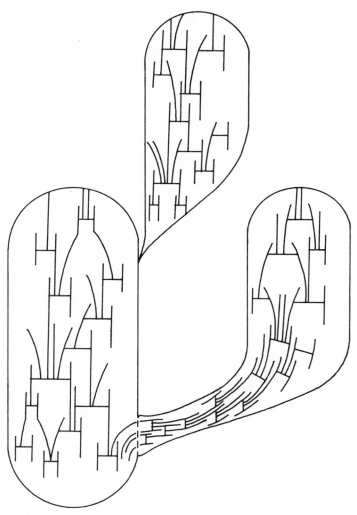

Figure 3.1 The view of the nature of species of sexually reproducing organisms adopted in this book. Three species are
depicted as discrete clusters demonstrating a within-group
pattern of *parental* ancestry and descent. The three species
also exhibit an among-group pattern of *phylogenetic* ancestry
and descent, with one species serving as an ancestor for the
other two in this particular example. Thus species are individuals, with their own internal cohesion, and give rise to descendant individuals rather like a parental *Hydra* buds off
young individuals.

Figure 3.1 illustrates our basic concept of species. They may be pictured as being somewhat like individuals of *Hydra* spp. (freshwater hydrozoan coelenterates). Each individual possesses its own identity and integrity. Occasionally new descendant individuals appear by budding, eventually detaching from the parent *Hydra* and establishing themselves as fully discrete descendant individuals.

We can summarize the foregoing into a simple definition of species: *a species is a diagnosable cluster of individuals within which there is a parental pattern of ancestry and descent, beyond which there is not, and which exhibits a pattern of phylogenetic ancestry and descent among units of like kind.*

Other Definitions of Species

Species have long been recognized as coherent units in nature. In an evolutionary context, biologists familiar with species tended to equate evolution with the origin of species. For example, Darwin's 1859 monograph on organic evolution, which is basically a treatise on adaptation and selection as an argument for the validity of the very notion of evolution, was nonetheless entitled "On the Origin of Species," even though relatively little explicit attention was paid to modes of origin of these reproductive groups. Appreciation of the role species play in evolution, however muddled it has seemed at times,[1] nonetheless has remained sufficiently clear that all so-called "biological" or "evolutionary" definitions of species have centered upon their nature as reproductive communities. For instance, Mayr's (1940; 1942:120) definition states that species are "groups of actually or potentially interbreeding populations reproductively isolated from

1. A major theme of this book, as developed at length in chapter 6, is that many of the key architects of the "synthetic theory" (mostly geneticists and paleontologists) have overlooked the significance of the existence of species as *discrete entities in time as well as space* and have therefore produced a rather curious body of theory. Nonetheless, the importance of species (reproductive communities) in the evolutionary process has always been sufficiently clear to the most ardent "transformationist" that all modern definitions of species from an evolutionary point of view have stressed the reproductive community aspect of species. For a fuller explication of our views on contemporary evolutionary theory, see chapter 6.

other such groups." Later, Mayr emended the definition (1969:26): "Species are groups of interbreeding natural populations reproductively isolated from other such groups." In removing the phrase "actually or potentially," Mayr avoided the pragmatic difficulties in evaluating "potential" interbreeding.

Simpson (1951) proposed an "evolutionary" species definition, subsequently modified slightly to read: "An evolutionary species is a lineage (an ancestral-descendant sequence of populations) evolving separately from others and with its own unitary evolutionary role and tendencies" (Simpson 1961:153). This lineage concept of species emphasizes reproductive continuity through time. Wiley (1978) has recently reviewed various species concepts, concluding that Simpson's seems to agree better with patterns in nature. The definitions of both Mayr and Simpson cited above are perfectly consistent with the definition we have adopted. All center around species as natural and discrete reproductive units.

It is to be stressed at this juncture that these and other, similar definitions of species are concepts, pictures of patterns in nature. Mayr's deletion of "potentially" from his earlier definition points out an ironic, if not amusing, pseudo-problem with the species concept: it was repeatedly pointed out that it is realistically impossible to assess whether or not individuals living in the remote corners of a far-flung species' range could in fact mate successfully with one another. Similarly, paleontologists have spent a great deal of time worrying if, within a hypothetical lineage persisting, say, 1 million years, individuals within that lineage a million years apart could have successfully interbred. How far back in time within the lineage *Homo sapiens* could modern individuals mate with their predecessors? Such questions seem ludicrous. As pragmatic problems in species recognition (see Sokal and Crovello 1970, for a critical appraisal of the "biological species concept" from a pragmatic standpoint), such problems are not laughable. But the *concept* of reproductive continuity is another matter, related to but not the same as that of species recognition. Thus allopatric (geographically separated) individuals need not be able to mate in order to belong to the same species. The important issue, insofar as the concept is concerned, is a plexus of reproduction across the entire distribution of a species. Similarly, the concept demands merely an unbroken plexus of parental ancestry and descent through time, and *not* that the end members, separated by our hypothetical 1 million years, could have successfully interbred given the opportunity.

Discontent with the pragmatic problems of recognizing reproductive units in nature (discussed more extensively below), has resulted in the invention of other sorts of species concepts. These various sorts of species have been ably reviewed by Cain (1954). The concepts are unified, seemingly, solely by the desire of the exponents to find an easy way of delineating basal taxonomic groups. Their eschewal of an evolutionary connotation robs them of any theoretical biological interest. Consequently, we shall not review them further here.

Recognition of Species

The recognition of reproductive units is sufficiently difficult in practice as to have led many professional taxonomists to ignore the so-called "biological species concept." Our position is that, whereas a concept difficult to apply in practice is not thereby *ipso facto* a poor descriptor of actual patterns in nature, a species concept should nonetheless contain sufficient elements to allow predictions about the nature of species useful in practical species recognition.

The essence of the problem of species recognition is the simultaneous (a) elimination of all sampled organisms beyond the limits of a biological species, and (b) inclusion of all those sampled elements properly belonging to that species. We want neither to over- nor to underestimate the actual number of species represented in a sample of organisms; we also wish to minimize the misclassification, or misallocation, of organisms to particular species.

Furthermore, delineation of species can—and properly should—be regarded as a hypothetico-deductive operation. The composition of a species and determination of its limits constitutes a hypothesis. The systematist's problem is to reject all but the least unlikely hypothesis of species composition among the organisms being considered. Evaluation of conflicting hypotheses of species composition depends upon a comparison of the predictions about species' properties arising from the definition, with the patterns of intrinsic and extrinsic properties displayed in the material at hand.

What are the general predictions arising from the definition of the species concept we have adopted? There are two components to our definition: (a) within-species reproductive cohesion (parental ancestry and descent), and (b) among-species phylogenetic ancestry and descent. The first of these components gives rise to the general criterion (direct demonstration of reproductive cohesion) and three ancillary predictions: Species should exhibit (a) generally continuous distributions of intrinsic properties within species (figure 3.2), (b) continuous distributions in space (figure 3.3), and (c) continuous distributions in time (figure 3.3). All three of these latter predictions are relatively weak; exceptions to the first two occur frequently, and the third relies on notoriously spotty information from fossils. Moreover, each has a circular aspect; simple demonstration of continuity of any or all three of these properties is no guarantee that representatives of more than one biological species have not been included.

The second (phylogenetic) component of the species definition leads to a different set of predictions concerning the pattern of distribution of characters. The full strategy for recognizing species in a hypothetico-deductive framework includes both sets of procedures. First we discuss the general reproductive criterion and its ancillary three predictions.

Evaluation of Hypotheses of Species Identity: The Reproductive Criterion

The first step in the analysis of species composition is the segregation of the organisms sampled into "piles" of similar or nearly identical organisms, such that no further subdivisions (clusters) are evident. Thus the number of species actually represented is likely to be overestimated. For example, there may be two piles, one of males and one of females, which may eventually be shown to be sexual dimorphs of the same species. Alternatively, the number of species may be underestimated; sibling species (see Mayr 1963, chapter 3; Dobzhansky 1951:267 ff.) are closely similar and closely related true species (i.e., reproductively isolated units) which are, nonetheless, so similar as to be practically indistinguishable. The point here is that species must be diagnosable. If the systematist fails to notice

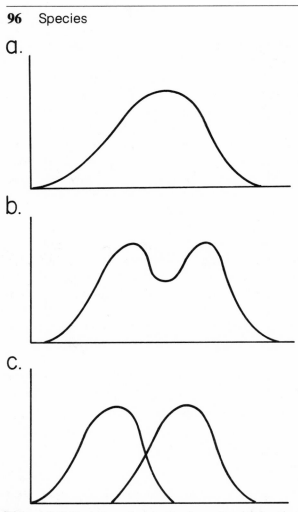

Figure 3.2 Variation of a single character. (a) A unimodal distribution within a single species. (b) Bimo-
dal variation (dimorphism) within a single species.
(c) Unimodal distribution within two separate species. Patterns of variation depicted in b and c are
difficult, if not impossible, to distinguish and thus are
of limited use as evidence of species composition.

the slight differences between two closely similar species, the error
is carried along throughout the analysis. There is no automatic way
to detect sibling species if they are not diagnosed as different at the
outset. Only additional data in subsequent analyses are likely to
reveal this error.

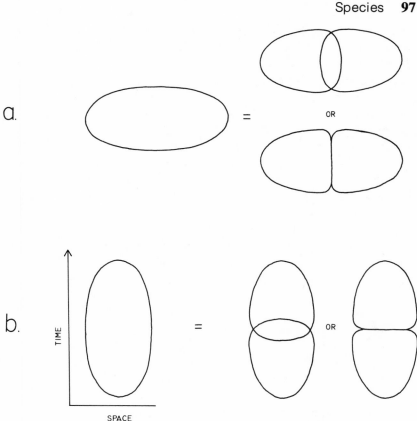

Figure 3.3 Problems in use of distributional data (extrinsic properties) to delimit range of a biological species. (a) Geographic distribution. The range of the hypothesized species may actually represent the combined ranges of two separate species either partially sympatric or wholly parapatric or sympatric. (b) Temporal distribution. The range of a putative species may actually represent the combined ranges of two separate species which either overlap or are mutually exclusive (i.e., time-successive).

Hypotheses of species composition are therefore various combinations of the initial discrete diagnosable clusters discriminated at the outset of the analysis. The reproductive criterion—direct demonstration of reproductive cohesion—comes into play first. For instance, there may have been two separate samples, not initially considered the "same," which may turn out to be males and females. Sexual dimorphism is a common characteristic of many different groups of sexually reproducing organisms. Morton (1958:139) discusses an extreme example. In deep-sea cephalopods of the genus

Argonauta, sexual dimorphism is pronounced. The large female carries a "paper" shell as an egg case, while the tiny male (only about an inch long) is reduced essentially to a reproductive organ, and thus not immediately recognizable as an argonaut in terms of its superficial morphological features. To confuse the issue even further, the male reproductive organ detaches and moves about freely within the mantle cavity of the female. The detached organ was long regarded as a parasite, even by Cuvier who, according to Morton, dubbed it *Hectocotylus octopodis*—which is doubly bizarre, as the generic name is the anatomical term for the male reproductive organ in cephalopods! Only when the full anatomical details of the males became known were there ample grounds for the hypothesis that these two quite dissimilar groups should be united into a single, albeit highly dimorphic, species. Other, less extreme cases, where reproduction has been observed (as in birds of paradise) have resulted in the synonymizing of two "species" into one—in other words, the recognition of two or more clusters as being the "same" and so naming them. In general, direct demonstration of interbreeding—preferably in the wild, but also in the laboratory—as confirmation of the general reproductive criterion of the species concept we have adopted, is the most direct means of uniting two or more previously separated taxa into a firm hypothesis of the identity of a biological species.

In the absence of direct information on reproductive behavior, some patterns of bimodality will nevertheless suggest the presence of sexual dimorphism. In recent years, some systematists working with ammonites (externally-shelled fossil cephalopods) have asserted that, in cases where two closely related and similar "species," which are found together in the same rock units, differ only in (a) relative size, (b) size of the initial chamber, and (c) presence or absence of a lappet (an armlike projection of the shell margin), they were dealing with a single dimorphic biological species. The smaller shells, bearing lappets, are hypothesized to be the males, by analogy with the known morphological characteristics of dimorphism in living cephalopods. However persuasive these cases may be, the formulation of a hypothesis of sexual dimorphism based on these kinds of data is actually based solely on inference from morphology and can only be evaluated further with the phylogenetic criterion, as we develop below.

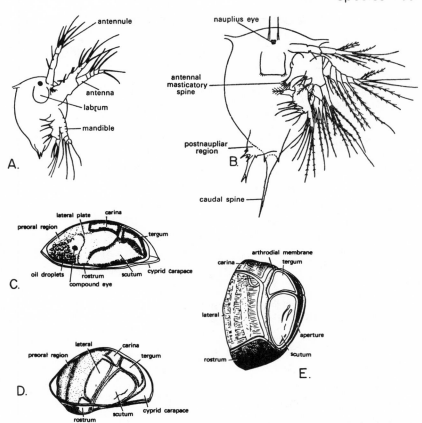

Figure 3.4 Larval development and metamorphosis of the acorn barnacle, *Balanus balanoides*. (a,b) First- and fifth-stage nauplius. (From Bassindale (1936), from Anderson, 1973, figure 117.) (c–e) Metamorphosis of (c) attached cyprid stage to (e) early adult stage. (From Runnström from Kaestner 1970, figure 9–15, with permission of Gustav Fischer Verlag.)

The reproductive criterion of parentage also comes directly into play when different stages in the life cycle of individuals are established by direct observation and experiment. The life cycle of barnacles (figure 3.4) includes several discrete larval stages. Early in ontogeny the larvae are free-swimming, and thus more typically crustaceanlike than the rooted, or cemented, adult stage. Much research in systematics, particularly in the various parasite groups, is devoted to the direct demonstration that larvae found either free-living or parasitizing another organism are the juveniles of an adult form parasitizing another, wholly different kind of organism. This is another way in which demonstration of parentage can reduce the number of clusters of discrete taxa into hypotheses of the existence of a single species.

Again, inference of continuity in the dissimilar ontogenetic stages of fossil organisms, or those elements of the Recent biota studied only from preserved materials, becomes a matter of the evaluation of intrinsic properties, and is not the same as the direct observation of such ontogenetic transformation. However, cases abound in which larval stages have been plausibly linked to adult stages on the basis of (a) associational (distributional) evidence, or (b) the more compelling demonstration of a complete series of morphological intermediates. Such demonstration is only possible within groups where progression between ontogenetic stages is gradual (e.g., in echinoderms such as edrioasteroids; see figure 3.5) and does not involve metamorphosis from one discrete stage into another, as in holometabolous insects and many parasites.

The reproductive criterion also comes into play directly in instances of hybridization. For present purposes, we can recognize two distinct aspects of hybridization: (a) occasional hybrids between two species, in which case the hybrids do not represent a taxon in and of themselves, and (b) situations where a new species has been formed by hybridization between two pre-existing species. Only the former consideration is relevant in terms of the general problem of species recognition. Again, direct field and controlled laboratory experimental evidence is frequently adduced, demonstrating the occasional hybridization between two species otherwise considered distinct. For example, hybrids between the American toad (*Bufo americanus*) and Fowler's toad (*Bufo fowleri*) are occasionally encountered, and can be raised in the laboratory. But again, putative hybrids, (i.e., specimens which appear to be intermediate between two taxa otherwise considered discrete biological species) are merely suggestive. What is the significance of such observations? How rampant must hybridization be before two clusters, formerly considered discrete reproductive communities, are merged as a single biological species? We enter a gray area here: If the hybrids are fully interfertile with both parental "species," we reject the hypothesis that these "species" are in fact discrete. Any sign of reproductive inviability within the hybrids, whether among themselves or with

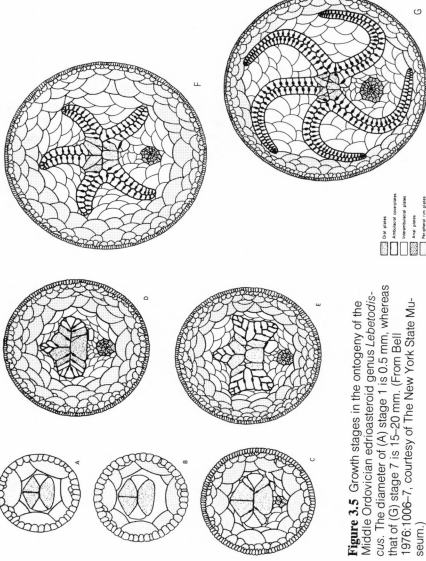

Figure 3.5 Growth stages in the ontogeny of the Middle Ordovician edrioasteroid genus *Lebetodiscus*. The diameter of (A) stage 1 is 0.5 mm, whereas that of (G) stage 7 is 15–20 mm. (From Bell 1976:1006–7, courtesy of The New York State Museum.)

Oral plates
Ambulacral coverplates
Interambulacral plates
Anal plates
Peripheral rim plates

either of the two "species," is presumptive evidence that the two clusters indeed constitute two separate reproductive communities, coherent within themselves but imperfectly separated from each other. Cases where only morphological intermediates are observed, with no concomitant breeding data, are incapable of decisive resolution, and further evaluation can only come from analysis based on the phylogenetic criterion.

In organisms reproducing wholly asexually, direct observation also confirms patterns of parentage, but in any case would appear to be of little use in reducing the number of putative diagnosable taxa in the manner described above for sexually reproducing organisms. There are grounds for doubting whether any organisms are strictly asexual, however, in which case the application of the parentage criterion from the species concept we have adopted is generally valid. As long as occasional sexual reproduction occurs (as in, for example, Foraminiferida), the parental pattern of ancestry and descent component of the species concept applies.

Thus the reproductive criterion, to the extent that parentage can be actually observed, is useful in the direct merging of separate, diagnosable clusters of organisms into single taxa (biological species) in cases of (a) sexual dimorphism, (b) disparate morphological stages in ontogeny, and (c) in some instances of hybridization. In the absence of direct evidence, the reproductive criterion may still be suggestive of such hypotheses in those instances where both extrinsic and intrinsic features conform to known patterns in related groups, but only as an inference much in need of further test.

Apart from direct observations of parentage, the three ancillary predictions from the species concept (continuity in distribution of intrinsic properties among individuals and continuity of distribution of individuals within the species in space and in time) offer further means of uniting previously sorted population samples in order to evaluate various hypotheses of species identity and composition.

Use of intrinsic properties—observable traits of individuals—at this stage depends upon the nature of the subdivision of the basic units in the first place. If the clusters of specimens were originally sorted because they seemed to be recognizably different, diagnosable taxa, then recognition of a continuous distribution of a variable trait (figure 3.2), if such exists, would conflict with the already perceived pattern of discontinuous variation, and further analysis would have to include the search for synapomorphies. However, if the clusters to be united were originally sorted on some other criterion—e.g., different localities in space and time—then such general continuity of intrinsic properties, including identity among two or more samples, arising as a prediction from the notion of parentage, may be used to unite clusters as a hypothesis of species identity.

Further preliminary merging of clusters can be effected by uniting clusters with contiguous distributions. If clusters were originally delineated primarily on the basis of their provenance from localities separated in space and time, and provided that the general aspect of the intrinsic properties suggests continuity among samples, then distributional contiguity (i.e., conformance to either of the two predictions of continuity in extrinsic properties) may be used to merge clusters as a hypothesis of the existence of a biological species (figure 3.3). Samples of organisms which appear to be nearly the "same"—i.e., the within-sample variation is nearly as great as the among-sample variation—and which, when mapped, show a continuous or quasi-continuous spatial distribution, may be hypothesized to be drawn from a single species. (See figure 4.12 for an example.)

This sort of approach, while arising as a prediction from the definition of biological species, and by conforming in a general way to the perception of intrinsic properties, nevertheless suffers from two basic flaws: (a) the negative use of the parentage criterion from the definition, and (b) the inability to specify limits, at least in more complicated cases, to the distribution. In other words, there remains the problem of either including too many or too few samples in an identified species.

Allopatric ("of another country") taxa display a geographic distribution wholly exclusive of one another. Sympatric ("of the same country") taxa occupy the same, or at least overlapping, ranges, such that individuals belonging to these taxa come into direct contact. Parapatric distributions involve close approximations of the limits of ranges of essentially allopatric taxa. Although species may occupy the same geographic area, they may in fact occur in different habitats within that area and thus not be truly sympatric, and therefore not have the opportunity for interbreeding. Lions and tigers are fully capable of interbreeding, yet they do not occur in the same habitats in India and therefore do not interbreed in nature. Allopatric distributions—cases where there are disjunct distributions of taxa which may or may not belong to the same species—are numerous. In such instances, obviously, there cannot be any positive evidence for interbreeding (parentage), and the point remains moot, on distributional criteria, as to whether or not one or two species are actually represented by the samples. In sympatric distributions, a pattern of parentage may be observed within each of two units hypothesized to represent distinct species, but the absence of evidence for parentage between the two "species" is negative evidence; the hypothesis that the two putative species are in fact distinct, true species stands until rejected by evidence of interbreeding.

The third and final ancillary prediction of the parentage component of the species definition is continuity in temporal distribution. By definition, such parental continuity must remain unbroken through time. Temporal distributions, naturally, can only be studied empirically with fossils. There are few circumstances in the fossil record where distributions are thoroughly continuous for any hypothesized species except for relatively short intervals in the total duration of a hypothesized species. The only exceptions are for species whose temporal durations are extremely short; a species may be known from only a single occurrence, in which case, for all practical purposes, it is not known to have a non-trivial temporal distribution. In all other instances, vertical gaps occur in ranges of taxa, and the inference is made that (1) the species was living there but individuals were not preserved, or (2) the species was living elsewhere during that period of time, when the environment was perhaps inimical to the species. In other words, the species remained in existence, but occurred elsewhere. But what limits can be drawn to such inferences? A paleontologist might be willing to "allow" a gap of a few thousand, or even perhaps a few million, years, but would probably be unwilling to hypothesize that two samples, one from the Middle Jurassic and the other from the Upper Cretaceous, with no known intervening samples and otherwise appearing to be exactly the same, would actually represent the existence of a single species. Such a temporal gap (some 100 million years) would seem to be too long— yet it must be admitted that it might be possible for a species to exist that long. Again, negative evidence of this sort is simply not a conclusive means of testing a hypothesis of species composition.

We conclude that the basic proposition of the biological species concept—continuity of parental ancestry and descent within but not without the species' limits—serves as positive evidence for the merging of diagnosable clusters into species. A systematist may reject the hypothesis that two samples are drawn from two discrete species, if it can be shown that they are males and females regularly interbreeding in nature. But failure to demonstrate such a pattern of reproductive cohesion does not reject the hypothesis that the two samples constitute a single species, as there are too many sources of experimental and observational error, as well as problems of allopatric and allochronic distribution, which could account for such a failure. Likewise the three ancillary predictions arising from the parental component of the species definition offer only weak tests of a hypothesis that any two samples may actually represent a single biological species. Negative evidence, whereby samples show neither unimodal distribution of intrinsic properties, nor continuous distributions in space and time cannot constitute a particularly strong rejection of the hypothesis that any two samples represent a single biological species. Clearly, there is a need for a stronger means of testing hypotheses of species compositions. Such a test must be consistent with the parental criterion and the evidence adduced for the three predictions arising from this criterion. But it must be independent. As such it must arise from a wholly different prediction from the definition. The only way for this to be effected is to look to the other component of the species definition we have adopted earlier.

Evaluation of Hypotheses of Species Composition: The Phylogenetic Criterion

With the exception of sibling species, the procedure for forming and evaluating hypotheses of species composition under the reproductive criterion initially overestimates the number of species present in a sample of organisms. In contrast, as we showed earlier in this chapter, hypotheses of species composition based on distributions of synapomorphies (one or more autapomorphies for a minimally defined monophyletic group) underestimate the number of species present. This discrepancy arises from the status of species as evolutionary units. Some species, perhaps including some represented by the samples at hand, are ancestral to others. We do not know whether ancestors are present in the sample or not.[2] The possibility that some of the species within a sample may be ancestral to other species (whether or not the direct descendants themselves are included in the sample) implies the possibility that not all species in a sample will be characterized by its own set of synapomorphies. Thus a straightforward application of the procedures of cladistic analysis, which evaluates hypotheses of synapomorphy and hence hypotheses of composition of monophyletic taxa, may not be sufficient to recognize all species represented in a sample. It would tend to underestimate the number of species actually present.

There is one case in which the phylogenetic procedures— search for patterns of synapomorphy—may *overestimate* the actual number of species represented in a given sample of organisms. This case involves a species morphologically well differentiated over its geographic range. A feature of most models of allopatric speciation is that morphological differentiation (i.e., development of evolutionary novelties) may occur within the geographic range of a species prior to the onset of full geographic (and hence reproductive) isolation among the populations. Thus, by the reproductive criterion, there is one species. But, by phylogenetic analysis, specifiable clusters, complete with autapomorphies, would indicate the presence of two (or more) species.

Such cases commonly arise in the analysis of allopatric or parapatric distributions of distinctive populations which themselves display some morphological homogeneity. In the hypothetical example, it was assumed that reproductive isolation was known *not* to have occurred. In reality, the number of species present in such a situation must remain moot, as earlier noted.

Nonetheless, terminal species, as well as species without any descendants (no matter how remote) within the sample being considered, will be diagnosable in terms of uniquely derived characters, or autapomorphies. This consideration suggests a general procedure: patterns of similarity (synapomorphy) should be evaluated according to the criteria developed in the preceding chapter on cladogram analysis. Those organisms which do not cluster into piles

2. As will be developed in our discussion of patterns of speciation, ancestral species can live on, coeval with and perhaps even outliving, their descendants. Thus fossils are not required for ancestors to be present in a sample.

based on synapomorphy will, therefore, appear relatively plesiomorphous in all features considered. According to the criteria developed in the preceding chapter, it is impossible to postulate the existence of a (monophyletic) taxon based solely on plesiomorphous similarities. Yet, as we have seen, species may not conform to the definition of monophyly. The residual organisms are diagnosable, presumably, only in terms of the absence of apomorphies, or, put in a more positive fashion, the unique retention of plesiomorphies. It is to be noted that we are not here considering whether or not a species is an ancestor (see chapter 4). We are concerned, rather, with the problem of recognizing species, which might happen to be ancestors, as species in the first place.

Further evaluation of the identity of collections of such plesiomorphous organisms as species can only come from a comparison of the clusters resulting from phylogenetic analysis with those suggested by the reproductive criterion and its three ancillary predictions. The procedure in the latter case initially overestimates the actual number of species, whereas in the former phylogenetic procedure, the number is generally underestimated. Should the plesiomorphous clusters resulting from the phylogenetic procedure appear cohesive under the reproductive criterion as outlined above, the hypothesis that the plesiomorphous cluster constitutes a single coherent species should be provisionally accepted. The hypothesis can always be rejected by showing that a portion of the species so identified is more closely related to (shares synapomorphies with) some other group or is itself autapomorphous and best regarded as a distinct species.

In a situation in which one sample is plesiomorphous with respect to a second—and thus possibly its ancestor—we are methodologically forced to recognize two discrete species where, in fact, there might only be one. If the two species are contemporaneous, there is no confusion. We recognize two species and then consider, according to the procedures set forth in chapter 4, whether the plesiomorphous species was in fact ancestral. But if the first, plesiomorphous species is exclusively known from an earlier period of time than that of the relatively apomorphic species, the possibility arises that the two "species" are samples from a single reproductive continuum. Should this be the case, under the species definition adopted here, the two samples would be conspecific. Methodologically, however, there would be no way to establish the (former) existence of the continuum. We would be forced to continue to recognize two distinct species.

Analysis of Species Composition: A Summary

We have developed two approaches to formulating and testing hypotheses of species composition. Each stems from a component of the definition of species adopted in this book. The reproductive cohesion aspect or component of the definition initially tends to overestimate the actual number of species. The number of diagnosable clusters is reduced primarily by direct demonstration of patterns of parental ancestry and descent; hypotheses are further evaluated by reference to predictions of continuity in intrinsic and extrinsic properties of species. The latter predictions are, however, weak.

The second approach is a variant of cladistic analysis. Species ancestral to other species *in the sample* cannot, by definition, possess a uniquely derived set of characters. We have advocated that clusters of plesiomorphous individual organisms can be hypothesized to constitute a biological species. In such a case, agreement with the reproductive criterion is particularly crucial. The hypotheses can be rejected with additional information showing that some of the individuals allocated to the hypothesized plesiomorphous species are either related to some other cluster, or are members of another unique group.

Species Recognition in Practice: Some Examples

Case 1. New material Platnick and Shadab (1976) have recently reviewed the spider genus *Zimiromus*. In so doing, they have described some 17 new species (i.e., species which had never previously been recognized in the taxonomic literature). Their general procedure was as follows: collections were procured (newly collected field samples of museum specimens) and prepared and examined in the laboratory. Specimens were recognized as showing synapomorphies with other species of the genus *Zimiromus*, i.e., they conform to the synapomorphies in the diagnosis of the genus. Such allocation permitted comparison with other species of the genus, which was essential for the evaluation of the distribution of characters within the samples at hand (because a higher-level hypothesis must be accepted prior to analysis to permit outgroup comparison; see chapter 2).

Subsets of the specimens were further recognized as being, in some ways, unique, i.e., they were different in some specifiable way from all other known species within the genus. They were thus diagnosable, and the characters utilized for the diagnosis were unique to the group, but not necessarily apomorphous. For instance, the two species *Z. penai* and *Z. brachet* (both from Ecuador) were distinguished solely by the relative width of the median epigynal ducts (a feature of the reproductive tract in females). One state may well be plesiomorphous, the other apomorphous. What matters is that a character be unique to a (putative) species; a uniquely retained plesiomorphy is as valid a criterion of a species as a uniquely derived feature.

In instances in which both males and females are known (e.g., the description of *Z. jamaicensis*), there were two initial clusters. Inasmuch as spiders in general are known to reproduce sexually and males and females of *Z. jamaicensis* share a large number of intrinsic features, the procedure was simply to include both males and females in the species, even in the absence of any direct information on mating behavior.

Platnick and Shadab (1976) also described *Z. bimini* from South Bimini, in the Bahama Islands. The males of this species share an apomorphous condition with males of *Z. jamaicensis* (the presence of single, long retrolateral apophyses). Thus the authors recognized a nested set of synapomorphies between the level of the genus and that of its discrete, included species. Why then were these two species not simply recognized as comprising a single (by virtue of the synapomorphy) albeit variable (by virtue of the diagnosable differences) species? In point of fact, were the pattern of parental ancestry and descent known in detail for both groups, it is conceivable that the pattern would be shared by the two (despite the rather large geographic distance between the two known samples) and that thus they would constitute a single species. But absence of any such direct information left the systematists no choice: they clustered the diagnosable units within the spiders sampled (i.e., males and females, etc.) on presumed parental criteria, consistent with the modified version of cladistic analysis that we have presented.

Case 2. Revision of species Much of the work of the practicing systematist lies in the revision of hypotheses of species composition already present in the literature. Rearrangement of hypotheses of species composition has two aspects. One is the removal of some samples previously allocated to a certain species. In such a case, the samples so removed are referred to one or more other species, either already described, or described as "new." In so doing, the concept of variation of the intrinsic or extrinsic (or both) properties of the species is decreased. The second aspect is the inclusion of additional material (newly found, in which case the range of variability of intrinsic and extrinsic features of a species is expanded), or two species are found not to be separate, diagnosable groups, as previously hypothesized in the literature. In the latter case, the species are *synonymized,* i.e., are formally merged, and the name chosen follows the Rules of Zoological Nomenclature. A typical species revision may encompass both aspects: inclusion and elimination, in the refinement of a hypothesis of species composition.

As an example of simultaneous exclusion from one notion of species composition, with concomitant inclusion in another, in revisionary work, Eldredge (1972) studied a large sample of available specimens of the trilobite genus *Phacops* in (Upper) Middle Devonian rocks of eastern and central North America (figure 3.6). Several species had been previously described. All but two had been synonymized by earlier workers. Eldredge examined the type material for all described species and concluded that, on the basis of the examination of the morphology (i.e., solely on the basis of intrinsic properties), the previous synonymies appeared correct: some of the previously described species appeared indistinguishable from each other as diagnosable taxa. There seemed to be only two basically separate, diagnosable clusters of specimens, conforming to the described species *Phacops rana* and *Phacops iowensis.* One of these, *Phacops rana,* exhibited obvious subclusters, whereas *P. iowensis* showed far less variability.

Examination of samples, particularly from the Michigan Basin, revealed that many of the *Phacops* specimens previously identified

a.   b.

Figure 3.6 Comparison of (a) *Phacops rana rana* and
(b) *Phacops iowensis*. Though similar in general appearance,
the two species differ in a number of specifiable attributes,
including the number of columns of lenses in the eye.

as *P. rana* shared instead those general features of *P. iowensis* most
critical in distinguishing that species from *P. rana* (e.g., number of
columns of lenses in the eye, and mode of development of the tuber-
cles on the external side of the dorsal cuticle). Again, whether these
different characters were primitive or derived was irrelevant to the
analysis of species composition. In fact, the sister-species of *P. rana*
appears to be a slightly older European and African species,
whereas *P. iowensis* apparently shares a pattern of synapomorphy
with older species in North America and elsewhere. The reclassified
specimens simply shared the attributes (intrinsic properties) of *P.
iowensis,* rather than those of *P. rana*. As formerly constituted, the
samples did not fit evenly into discrete, diagnosable groups.

The diagnosable subclusters of *P. rana* presented the familiar
problem of assignment of such clusters to the appropriate rank. In
one instance, two subclusters were nearly identical except in terms
of the total number of lenses in the eye; earlier systematists had suggested that these two diagnosable clusters represented a case of
dimorphism, presumably sexual. The early ontogeny of these two
groups is identical; in addition, sexual dimorphism in lens number is
known in other groups of aquatic arthropods (e.g., *Limulus polyphe-
mus,* the horseshoe crab). But the two taxa seldom cooccur in the
same bedding planes (which perhaps would suggest sympatry) and
occur alone in some regions. For this reason, Eldredge (1972) rejected the hypothesis of sexual dimorphism and did not merge the
two subclusters.

Each of the five diagnosable subclusters within *Phacops rana* was defined on the basis of unique intrinsic characters. When mapped, the distributions of each proved continuous (as nearly so as fossils are likely to be) in space and time (see figure 4.12). Eldredge chose to label each of these five subclusters as a "subspecies," but the point is clearly moot. Indeed, the procedures outlined and illustrated in this chapter suggest that each of the five "subspecies" should have been accorded status as actual discrete species. However, it makes little difference as the point necessarily will remain moot in the absence of direct information on reproductive patterns within and among the diagnosable taxa.

Chapter
4

Modes of Speciation and the Analysis of Phylogenetic Trees

THE GENERAL methodology for producing cladograms developed in chapter 2 requires only one general assumption about the evolutionary process: that evolution has occurred and has created nested sets of evolutionary novelties (synapomorphous characters). The consideration of the nature of species and their identification requires a closer examination of the evolutionary process. The actual units of evolution are the reproductive communities called species. That some species are ancestral to others necessitates a modification of normal cladistic procedures in the analysis of species composition (chapter 3). Evolution of new species and the development of evolutionary novelties are not one and the same thing. Though related, novelties can be attained without new species being formed and the converse is true as well. We now discuss the topic of phylogenetic trees, their nature and method of construction. And in considering trees we shall need to examine more closely the actual evolutionary patterns of the origin of new species.

We define a phylogenetic tree as a branching diagram that depicts actual patterns of ancestry and descent among a series of taxa. Thus trees are far more specific sorts of statements than cladograms. Cladograms depict nested sets of synapomorphies, thereby defining monophyletic groups and simultaneously presenting a hypothesis of the relationships among the taxa. Cladograms depict relationships in a relative way: taxon A and taxon B are more closely related to one another than either is to taxon C. For instance, the statement that cats and dogs (Felidae and Canidae) are more closely related to each other than either is to, say, seals (Pinnipedia) is an example of a cladistic hypothesis. Such statements can be made about any presumed monophyletic cluster, of whatever categorical rank.

Trees, in contrast, depict specific hypotheses as to the manner in which taxa are related. We have already concluded that, logically, only species can serve as evolutionary units. Taxa of higher rank are merely monophyletic aggregates of one or more species, and thus do not exist in the same sense as do species and cannot serve as ancestral or descendant units (see also Wiley 1979; for a contrasting view, see Bretsky 1979). Nonetheless, the literature is replete with phylogenetic trees depicting ancestor-descendant relationships among genera and taxa of even higher rank. Thus our conception of phylogenetic trees is more restrictive than the usual notion. Phylogenetic trees as understood here are branching diagrams showing patterns of ancestry and descent among species. In chapter 6, we shall discuss the use of highly corroborated phylogenetic trees in evolutionary theories of species formation. At this juncture, we briefly review generalizations on patterns of speciation to expose the possible sorts of trees that can be produced.

Patterns of Speciation

Accepting the concept of species as discrete reproductive units, it follows that speciation is the origin of new reproductive communities. Such an event can only result from a pattern of splitting, or budding off from, an ancestral species. Descendant species arise, by definition, from a portion of the ancestral species. The only alternative, the wholesale, complete phyletic transformation of one entire species into another, is inadmissable because there is no postulated disruption of the continuum of parental ancestry and descent in such a situation. Separately *diagnosed* species resulting from such phyletic transformation are rather to be regarded as end members of an unbroken continuum; as we have earlier remarked, such end member "species" might be diagnosed as two, but constitute a single reproductive lineage and thus constitute a single evolutionary entity. Our review of speciation will therefore focus on modes of origin of new, discrete reproductive communities from ancestral species.

Consideration of the modes of origin of new species is complicated by the general model of phyletic evolution. We briefly mentioned phyletic evolution in chapter 3 in discussing the problem of handling samples a million years apart within a hypothesized single lineage, unbroken in terms of parental ancestry and descent. The point here is that phyletic transformation of one or more characters within such a lineage is frequently alleged to have produced descendant morphologies so different that the systematist has no choice but to recognize two discrete species. From a pragmatic standpoint, an ancestral species evolves directly into a descendant (figure 4.1). Procedures for tree construction outlined later in this chapter offer a general means of distinguishing situations in which new species arise by splitting or "budding" from those representing phyletic change within a single species, a microevolutionary process culminating in "pseudospeciation." For classificatory purposes (see chapter 5), all diagnosable species, whatever their mode of origin, are treated the same.

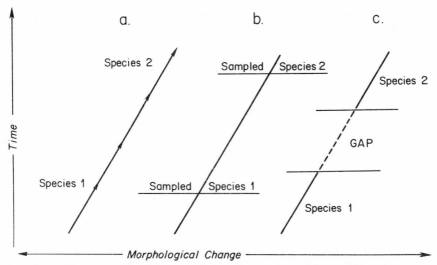

Figure 4.1 Aspects of the neo-Darwinian concept of the derivation of new species by transformation. (a) Transition from species 1 to species 2. (b) The transition is viewed as even and gradual, such that specifiable differences between species emerge only if samples widely separated in time are considered. (c) The incompleteness of the fossil record provides the gaps between samples, obviating the necessity for the systematist to divide the continuum in an arbitrary fashion.

Phyletic Transformation: Neo-Darwinism and Saltationism

As characterized in greater detail in chapter 6, models of phyletic evolution stem directly from notions of the transformation of intrinsic properties of organisms. They are not, at base, theories of speciation at all. The most common version of phyletic evolution in the English-speaking world is neo-Darwinian in nature. The essence of the neo-Darwinian model of phyletic evolution is that, over time, the genetic (hence biochemical, physiological, anatomical, and behavioral) attributes of species become modified such that, given the elapse of sufficient time, were the lineage to be sampled, it would be found to differ so much in its intrinsic properties that a systematist would recognize diagnosably different "species" (figure 4.1b). The theory emphasizes a completely continuous, evenly gradational and usually rather slow transformation of the intrinsic properties of the ancestral species into those of the descendant. Paleontologists and those they have influenced (see Cain 1954:111 for a lucid exposition of this model) have repeatedly stated that, were the fossil record perfect, systematists would be faced (even plagued) by the continuous series of forms that could only be broken up in the most arbitrary fashion and pigeonholed into the taxonomic hierarchy. Thus, from this point of view, we are told that it is fortunate indeed that the fossil record is so incomplete (figure 4.1c), such that the isolated glimpses afforded by sporadic preservation of samples of these otherwise continuous, gradational lineages allows systematists to recognize discrete taxa in a non-arbitrary fashion: the gaps in the fossil record provide the necessary arbitration.

We discuss the putative mechanisms underlying phyletic change in detail in chapter 6. For the moment it is only necessary to consider the evidence appropriate for the evaluation of this model as a general hypothesis of morphological change in the evolutionary process. Two independent bodies of evidence have been cited in support of the general notion of phyletic transformation as a major mode of evolutionary change: (a) extrapolations from experimental and theoretical (mathematical) genetics and (b) empirical evidence from the fossil record. Only the latter is directly concerned with pattern. The former is concerned with a hypothetical mechanism and has been ably characterized and defended by Simpson; we discuss it in greater detail in chapter 6. The model has been severely criticized as an unjustified extrapolation into an inappropriately much larger time dimension (Eldredge and Gould 1972, 1974; Gould and Eldredge 1977; also chapter 6). Of greater importance at this point is the empirical evidence of pattern: what is the evidence for (a) within-species gradual change through time, and (b) what is the evidence for successional occurrence of ancestors and descendants? The first kind of evidence is crucial to the neo-Darwinian theory of transformational speciation (but not to saltationism), and the second kind of evidence is crucial to the entire concept of transformational speciation.

The dearth of examples of continuous change within lineages, including change within segments designated as nominal species, in the fossil record has been recognized since Darwin's day (see especially Gould and Eldredge 1977, for a review of putative examples of phyletic evolution). Adherents to the neo-Darwinian theory of transformational speciation resort to the ad hoc hypothesis that the fossil record is too poor to reflect adequately this mode of evolution. And there is little reason to doubt that the fossil record is indeed spotty. Nonetheless, there are many cases where species, recognized according to techniques developed in chapter 3, have quasi-continuous ranges; in the case of marine invertebrates, these ranges may be 5–10 million years in duration, and in some instances even more. Within species delineated in this fashion, there is generally no net accumulated change in those intrinsic properties used by systematists to discriminate alleged ancestors from their supposed descendants. Most of the examples purporting to document such gradual transformation have since been rejected by subsequent workers.[1]

On the other hand, the more general proposition, i.e., that species succeed one another and do not display overlap in time (which is the more general prediction of the transformational hypothesis) remains a more open question. Indeed, many apparent ancestors

1. Many of the classic examples of phyletic gradualism in the paleontological literature are studies performed in Great Britain between 1899 and 1929. These include Rowe's study of Cretaceous echinoids (1899), Carruthers' (1910) study of the Carboniferous coral *"Zaphrentis" delanouei*, Trueman's (1922) paper on the Jurassic oyster *Gryphaea*, and Brinkmann's (1929) examination of the Jurassic ammonite *Kosmoceras*. Restudy of these and other cases has consistently failed to verify the gradualistic patterns of morphological change claimed by the original author. (Some cases have yet to be thoroughly restudied; see Gould and Eldredge 1977, for discussion and citation of subsequent reevaluations of these and other putative examples of phyletic gradualism in the paleontological literature.)

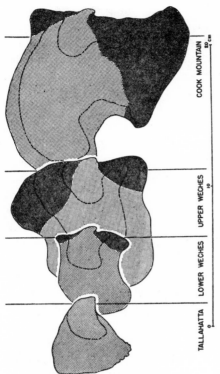

Figure 4.2 Stenzel's example of time-successive species hypothesized to form an ancestor–descendant sequence, and thus an example of derivation of diagnosable species by transformation. The upper three species are of the *Cubitostrea selliformis* stock, the lower the putative ancestor. Stenzel's explanation follows: "These are outlines of the left valves of large specimens of each species. They show the development of auricles and increase in shell size from species to species. The species are arranged with the earliest at the bottom and the latest at the top [N.B.: species are confined to the geologic formations named at the right side of the diagram—aus.]. Horizontal bars are the extensions of the hinge axes. Auricles are shaded differently than the main shell body. Some growth lines are indicated by dashed lines." (From Stenzel 1949:44, figure 8.)

seem to disappear from the record, succeeded by apparent descendants. These successively occurring taxa are recognized and characterized in the manner outlined in chapter 3. For example, Stenzel (1949; see figure 4.2) described a case in which four species, which he hypothesized formed an ancestor-descendant sequence, appeared to succeed one another in time, and in no instance was the ancestral species known to have survived as a contemporary of its descendant. Thus the general hypothesis of transformational speciation cannot be ruled out.

The most striking aspect of the neo-Darwinian (see chapter 6, p. 246, and footnote 1 for a discussion of the meanings of the terms "neo-Darwinian" and "synthetic") theory of transformational speciation is that it is not really a theory of speciation at all. Rather, it is a theory of change of intrinsic properties, effected primarily by natural selection working on a groundmass of genetic variation. Species emerge, at the end of an analysis, as semi-arbitrarily defined clusters of convenience. There is no qualitative difference alleged to differentiate species from genera and other higher taxa. Evolution proceeds "at the species level" but produces successive "species" that are all part of the same evolving lineage. Occasional splits in the lineage produce diversity, but genera and taxa of higher categorical rank are *ex post facto* summations of species-lineages that go on indefinitely. Species have no discrete limits in a phylogenetic sense; this view of the evolutionary process is, at base, incompatible with the view of species developed in the previous chapter and is not considered further until chapter 6.

"Saltationism" is less coherent than the neo-Darwinian view as a theory of transformational speciation. There have been several different mechanisms proposed for saltation. Moreover, neo-Darwinists have nearly unanimously viewed saltation as the only competing paradigm and have caused its nearly total eclipse in contemporary biology. Neo-Darwinists claim to have a mechanism (natural selection through geological time, as an analogue of experimental and artificial selection; see chapter 6), whereas most saltation theories have invoked mechanisms even more difficult to establish on an empirical (including experimental) basis.

However, saltationism—literally, the view that evolution proceeds by sudden jumps from one state to the next—never wholly abandoned the notion that taxa (species) do the jumping (figure 4.3). Thus at least most theories of saltation are theories of speciation.[2] Although most discussions of saltation by proponents and opponents alike seem to focus on transformational problems (e.g., explaining the origin of major structural differences among taxa within monophyletic groups), the creation of a "hopeful monster" directly implies the creation of a new species. Indeed, the only cogent arguments against particular theories of saltation appear to be those which see grave difficulties in creating a new, sexually reproducing species through the appearance of a single, individual hopeful monster. It is interesting, in this connection, that the term "mutation" was

2. In discussing saltationism as a general alternative to neo-Darwinian processes in the explanation of patterns of phyletic evolution, we do not wish to imply that saltational models have not also been concerned with splitting phenomena—speciation in the strict sense. Carson (1975), for example, writes of "saltational speciation." But much, and perhaps the bulk, of the older saltational literature deals with ancestral and descendant species occurring successively, in a phyletic sense.

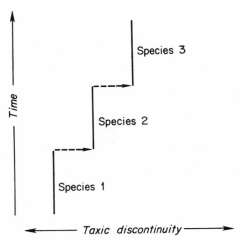

Figure 4.3 Speciation by transformational saltation. Each species is derived through the transformation of the ancestral species into its descendant. Transformations occur as sudden and discrete events (horizontal dashed lines), however, rather than by gradual, progressive change, the usual version of the neo-Darwinian model.

coined, not by a geneticist, but rather by the paleontologist Waagen (1869). And as Simpson (1944:48; 1953:81) has correctly pointed out, Waagen's mutations were taxa which, in a taxonomic context he called "varieties," but which seem to conform to the criteria of species recognition discussed in chapter 3. Waagen's mutations involved a concept of time-successive (i.e., nonoverlapping temporal ranges) monophyletic taxa arranged in a direct ancestor-descendant sequence. As previously acknowledged, the question as to the occurrence and relative frequency of such patterns remains open, insofar as the analysis of the fossil record is concerned. There is at least no evidence that clearly and decisively refutes the general proposition of saltation as a pattern of morphologically discrete, clearly differentiated successive species related in an ancestral-descendant fashion. Any particular model of saltation proposing a specific set of mechanisms should be experimentally testable, and it is strictly on this basis that saltationism would appear eventually to stand or fall, unless the general pattern predicted to occur in the fossil record is itself ultimately shown not to exist.

Speciation by Splitting

If phyletic transformation, whether guided by natural selection or through some other agency, does not suffice as a model for the production of new reproductive communities, it follows that the only real mode of speciation is that of splitting, the budding off of a portion of a reproductive community to form a new, descendant unit. In contemporary biology the words "speciation" is actually taken to mean "formation of new species by splitting." The definition of species adopted in the preceding chapter virtually requires such a process for the formation of new species. In his succinct review of theories of speciation, Bush (1975) refers to three basic types of speciation: allopatric (with two basic variants), parapatric, and sympatric. All three involve the derivation of a descendant species by the splitting off of a portion of the ancestral species.

Use of the terms allopatric, parapatric, and sympatric as descriptive modifiers emphasizes a central unifying factor in contemporary speciation theory: geography plays a key role in speciation. Speciation, the fragmentation of an ancestral species into two or more different species, can happen in another place (allopatric), and adjacent places (parapatric), or in the same place (sympatric). Thus stages in the differentiation and development of new species can, at least theoretically, be mapped. In conjunction with the representation of phylogenetic histories by branching diagrams, the "mapability" aspect of the speciation process leads directly into the related field of historical biography, the study of the distributional history of the earth's biota. Nelson and Platnick (1980) have discussed the relationship between systematics and biogeography at length, and the diversity of methodological approaches to this field is well treated in a recent symposium volume on vicariance biogeography (Nelson and Rosen 1980).

The core of any viable speciation theory is a mechanism for the disruption of the within-species pattern of parental ancestry and descent. A phylogenetically descendant species becomes reproductively isolated from its ancestor. Allopatric distributions, particularly involving species of low vagility, offer a de facto disruption of the pattern of parentage: disjunct populations simply cannot share a cohesive pattern of parental ancestry and descent unless there are dispersal mechanisms that bring them into contact from time to time. However, such disjunct distributions offer no prima facie evidence that two (or more) species are involved (see chapter 3). Shear (1975) has described the occurrence of the phalangids (harvestmen) *Caddo agilis* and *Caddo pepperella* from Japan. These species are otherwise known only from eastern North America. But samples within each species from eastern North America and Japan are indistinguishable (in terms of the characters examined thus far) and therefore, according to the criteria developed in the preceding chapter, there is no basis for recognizing these disjunct allopatric occurrences as representing two separate species, inasmuch as they are not separately diagnosable. The problem of the lack of an immediate and tight correlation between (apparent) reproductive isolation on the one hand, and morphological divergence (attainment of novelties, diagnosability) on the other is raised. Geographic isolation is an important (some would say "necessary") cause of disruption of patterns of parental ancestry and descent. Such disruption is the *sine qua non* of speciation. But such disruption may not be reflected in intrinsic properties needed to determine the presence of two species rather than one (see Ehrlich and Raven 1969).

Nonetheless, from the earliest days of analysis of modes of speciation, it has been recognized that geographic isolation affords the simplest, most direct means of effecting reproductive isolation. Wagner (1869), Romanes (1886), and a number of other nineteenth-century biologists noted that apparently closely related distinct species tended to "replace" each other geographically as ecological "vicars" or "vicariants," suggesting that the formation of new species is primarily a matter of geographic differentiation followed by reproductive isolation. Mayr (see especially 1942, 1963, 1970) has marshaled a large amount of data from the primary literature and has concluded that allopatric speciation is the predominant mode of formation of new species, at least among sexually reproducing animal species. This view has come to prevail in recent years in contemporary biology.

Bush (1975) has made a heuristically useful distinction between two basic variants of the theory of allopatric speciation (figure 4.4). In Bush's (1975) type *a* allopatric speciation, an ancestral species of rather broad geographic distribution becomes subdivided into two or more daughter-species. In such cases, the geographic range is generally thought to become disrupted through some change in physical geography within the ancestor's range. The two sections

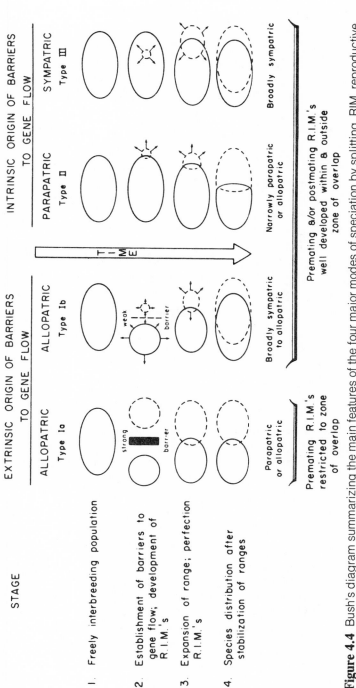

Figure 4.4 Bush's diagram summarizing the main features of the four major modes of speciation by splitting. RIM, reproductive isolating mechanism. (From Bush 1975:344, figure 1. Reproduced, by permission, from the Annual Review of Ecology and Systematics, volume 6 © 1975 by Annual Reviews Inc.)

simply gradually diverge—gene flow (i.e., mutual pattern of parentage) is automatically cut off, and each subunit gradually adapts (via natural selection) such that, should sympatry reoccur at some later point in time, the two subunits would be unable to reestablish a mutual pattern of parental ancestry and descent. They will have become totally reproductively isolated as an accidental byproduct of their physical separation. In such a model, natural selection is not envisioned as establishing reproductive isolation "deliberately" itself; rather, through selection and perhaps genetic drift, enough genetic differences accumulate to prevent the formation of viable hybrids should the opportunity subsequently arise. Or the reproductive behavior of the two units could have become sufficiently different in isolation so that hybrids will not be formed (see Dobzhansky 1951 for a thorough review of isolating mechanisms). The point to be stressed here is that, in isolation, the two units simply go their own way to a point where they cannot merge on the eventual onset of sympatry. Thus the model itself is a hybrid between the transformational view of phyletic change or "speciation" and splitting per se: intrinsic properties are gradually and slowly modified through time in two or more discrete but large populations that started as fragments of a widely distributed ancestral species. In cases where both fragments diverge perceptibly, such that each becomes diagnosably different with respect to the common ancestor (some situations in the fossil record might, at least theoretically, offer the possibility of recognizing this pattern), the pattern shown in figure 4.5a results. If one of the two fragments remains indistinguishable as a diagnosable taxon from the ancestral species, then the pattern of figure 4.5b emerges, which is indistinguishable from the basic pattern of splitting which results from the other model of allopatric speciation (type b of Bush 1975), as well as parapatric and sympatric speciation.

To the extent that the gradual divergence model of allopatric speciation can be corroborated (i.e., if it seems to fit the data in some cases better than alternative models), it affords perhaps the only means of preserving the transformational aspect of speciation without abandoning the concept that species are discrete evolutionary units. According to Bush (1975), this mode of allopatric speciation is a long-term process and is fairly common, especially among terrestrial vertebrates.

The second mode of allopatric speciation, the development of

a.

b.

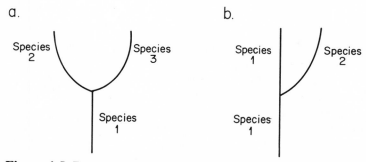

Figure 4.5 Expected consequences of various speciation models. (a) The expected outcome of type *a* allopatric speciation (Bush 1975). The two descendant species, derived from the splitting of the ancestral species, each develop autapomorphies. The ancestral species becomes "extinct by transformation." (b) The expected outcome of type *b* allopatric, parapatric, and sympatric speciation. The descendant species develops autapomorphies and the ancestral species persists unchanged.

peripherally isolated populations much smaller in size than the entire ancestral species, is, again according to Bush (1975:346) much more common than type *a* allopatric speciation. As in type *a,* selection in type *b* allopatric speciation is envisioned as effecting change in the adaptations of the isolated population, but smaller population size makes more rapid change possible. The expected result (figure 4.4) is a vicariant initial distribution of the ancestral species and its daughter species, which has become diagnosable by virtue of developing specifiable autapomorphies—evolutionary novelties which in this case are hypothesized to relate to adaptation, the occupation and exploitation of niche space. Thus the resultant pattern is as drawn in figure 4.5b, where a new, discrete species arises from an ancestor (which continues on unchanged, hence not diagnosable as a "new" species itself) in a relatively short span of time (geologically instantaneous). Moreover, the initial geographic pattern is vicariant, though, as Bush (1975) notes, there is ample opportunity for broad neosympatry, inasmuch as parent- and daughter-species are alleged to have somewhat different sets of adaptations, i.e., they occupy somewhat disparate niches.

As illustrated in figure 4.4., there are two fundamental forms of allopatric speciation on the one hand, and parapatric and sympatric speciation on the other hand. These latter two modes involve (a) an initial distribution, in which the daughter-species is directly adjacent to (parapatric) or wholly within (sympatric) the distribution of the parental species. Thus (b) selection for reproductive isolation is hypothesized to occur directly; i.e., reproductive isolation does not develop merely as an accidental by-product of divergence. The problem with such models has always been the difficulty of concocting a plausible explanation of how reproductive isolation can develop without the aid of extrinsic (specifically, geographic or physical environmental) barriers (see Mayr 1963 for an extended discussion). Bush (1975) has reviewed the possibilities and presented evidence that such processes can occur. Of direct interest to a systematist, however, is the fact that the main effect of both parapatric and sympatric speciation is identical to that of allopatric speciation involving peripherally isolated populations (i.e., type *b* allopatric speciation): the end result is a new species diagnosably different from the unmodified ancestral species. Thus type *b* allopatric, parapatric, and sympatric speciation all are predicted to result in the development of the pattern of figure 4.5b. Moreover, all three are held to produce new species rapidly. Thus, from the point of view of systematics, the only distinction to be drawn among these three models is the nature of the initial distributions of the ancestral and descendant species at the onset of reproductive isolation. If sympatric speciation is a real possibility (as Bush argues), then demonstration of sympatry between two closely related species does not automatically imply a time lapse sufficient to allow a change from an allopatric to a sympatric distribution. These considerations have implications primarily for historical biogeography. In terms of representing phylogenetic events on branching diagrams, there are no meaningful distinctions among these latter three modes of species formation by splitting.

Carson (1975) has recently discussed rapid speciation events involving peripherally isolated populations, stressing the possibility that such events can be so rapid as to appear "saltational" even in ecological time. When compared with the empirically established phenotypic stability evidenced by many minimally diagnosable taxa (i.e., hypothesized species) over truly long expanses of geological time, such rapid speciation events might indeed appear to be saltational. In most classic cases where saltation has been hypothesized, i.e., where entire species were judged to be transmogrified into descendant species by sudden jumps, by definition there can be no synchroneity of ancestor and descendant (i.e., no survival of the ancestral species as a contemporary of its descendant). Therein lies the only formal distinction, in terms of pattern only, between classic saltationism and speciation in the strict sense: in allopatric (type *b*), parapatric, and sympatric speciation, the expected pattern is synchronous overlap of ancestor and descendant, whereas the opposite is expected from classic saltation. In situations involving fossils, saltation can be rejected only by demonstrating temporal overlap between ancestor and descendant. Failure to demonstrate such contemporaneity is obviously negative evidence and thus not a strong test of successional or "classic" saltation. But, aside from this consideration, no amount of data pertaining to fossils is likely to allow a systematist to distinguish unequivocally among the possibilities of saltation, on the one hand, or speciation (whether type *b* allopatric, parapatric or sympatric), on the other. And there is the even more fundamental problem of the formulation and testing of hypotheses of phylogenetic ancestry and descent—phylogenetic trees—the subject to which we now turn.

Construction and Testing of Phylogenetic Trees

In this section we present a general strategy for the production and evaluation of phylogenetic trees. We shall develop the basic theme that trees have a relationship with cladograms such that, for any given problem, there are a number of possible trees consistent with any one cladogram (see also chapters 2 and 5 for further discussion of the concept that cladograms are sets of trees). The central problem of tree construction is to reject all but the least unlikely tree consistent with any given cladogram. As more specific statements, trees are in some ways more easily tested (i.e., rejected) than cladograms. Construction of trees, however, entails an additional set of assumptions not needed for cladogram analysis. Incorporating more information (e.g., on extrinsic properties of species), a tree is also further removed from the "data base" of pure character distributions among organisms than the background cladogram on which it is based. Thus trees probably can never be as highly corroborated as cladograms.

As diagrams purporting to show actual evolutionary events, trees portray hypothesized sequences of ancestry and descent. Given the corollary of any general notion of evolution, i.e., that such patterns of ancestry and descent must exist, the systematist can hardly be faulted for seeking them. And, as we shall see in chapter 6, highly corroborated trees are absolutely essential to the testing of competing theories of speciation. And it would be well to reemphasize at this juncture that only trees involving species as ancestors and descendants have any meaning beyond the cladogram level of analysis. There is no formal difference—just semantic confusion and the retention of non-monophyletic groups—between cladograms and trees involving taxa of rank higher than species.

The Topology of Trees

There are two different ways of drawing phylogenetic trees. Clarification of these topologies is essential to avoid ambiguity. In figure 4.6 we compare, purely from the standpoint of conventional notation, the various uses of "dots" and "lines" in cladograms and trees. In figure 4.6a, a typical cladogram is shown. Taxa are represented by dots; lines merely connect taxa to depict the hypothesized nested sets of synapomorphies. In figure 4.6b, we show a phylogenetic tree which is drawn with the same symbolic conventions as in the preceding cladogram: taxa (species) are dots, and lines depict a pattern of relationships, in this case one of ancestry and descent. In figure 4.6, we show the same tree, redrawn this time to show the known duration in time of each of the species. In this instance, solid lines stand for the entire known range of the species; hypothesized patterns of relationship are shown as dashed lines. The only "dot" is the unknown, hypothetical species X, the inferred common ancestor.

For ease of comparison, it is convenient to utilize the same notation in both the cladogram and its derivative trees. However, the notation of figure 4.6c, where taxa are represented by lines rather than by dots, is the style adopted in the overwhelming majority of trees published in the literature. Conceptually, it makes little difference, inasmuch as either system agrees with the fundamental notion of species as discrete individual entities in nature. For the remainder of this chapter, we shall adopt conventional symbolization, not only because it agrees with the bulk of past usage, but also because it

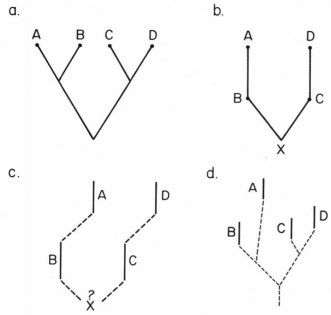

Figure 4.6 Comparison of cladograms and trees. (a) A cladogram. Species are represented by dots; lines connect dots to indicate nested patterns of synapomorphy. (b) A phylogenetic tree, one of several consistent with the cladogram of a. Notational symbols are exactly the same as in the cladogram; species are represented by dots, and the lines, in addition to indicating nested sets of synapomorphies, connect the species in ancestor–descendant fashion. The hypothesized ancestral species is represented by X. (c) A more conventionally rendered phylogenetic tree. In this diagram, species A, B, C, and D are depicted by solid lines indicating known temporal distributions. Dashed lines depict hypothesized lines of descent. The only "dots" used are for X, the unknown, hypothesized ancestral species. (d) The original cladogram redrawn using the notational conventions of the phylogenetic tree of c.

allows depiction of overlapping ranges among a series of ancestors and descendants. Such overlap is both practically and conceptually important. Pragmatically, different conceptions of speciation predict various different patterns of temporal distribution among ancestors and descendants; incorporation of known ranges in a tree therefore aids in evaluating the status of any particular species as an ancestor vs. a collateral descendant, and contributes as well to the evaluation of the relative merits of competing speciation models in any given instance. And conceptually, inclusion of time-range information on a branching diagram, be it a tree or a cladogram, is directly tied to notions of differential species survival, discussed in detail in chapter 6. As an example, in figure 4.6d, we redraw the cladogram of figure 4.6a using the conventions of the tree of figure 4.6c.

Construction of Trees: Additional Assumptions in the System

Construction of both cladograms and trees assumes a pattern of ancestry and descent connecting the species being analyzed. It follows as a natural goal of systematics to reconstruct that pattern in as fine a detail as possible—hence the zeal with which early evolutionary biologists converted the Scala Naturae into trees purporting to show actual evolutionary lines of descent. However, construction of trees, with ancestors and descendants specified among a series of species at the outset requires two additional assumptions not made with cladograms. The first assumption is that *the species actually involved in connecting speciation events are at least potentially present in the sample.* Another way to put it is that some of the species present in the sample may be the direct ancestors of other species under consideration. Cladograms, of course, require no such assumption. Cladograms depict nested sets of synapomorphies. In terms of relatedness of the taxa, a cladogram merely orders taxa in terms of nearness of common ancestry. Cladograms require no assumptions about the presence of ancestors in the sample (figure 4.7). Inasmuch as it is impossible to assume with certainty that direct ancestors are included in the sample, it appears that the construction of phylogenetic trees has an intrinsic flaw, requiring an assumption perhaps many systematists would not be willing to make. This accounts for the predilection for drawing trees among taxa of generic and higher rank (otherwise a logical absurdity, as already noted), which loosens the restrictions of this assumption.

Nonetheless, in certain circumstances, the assumption that actual speciation events are at least potentially included within a sample of species may appear reasonable. According to the majority of speciation models reviewed earlier in this chapter, ancestral species frequently survive as contemporaries of their descendants, and are apt to display characteristic geographical distribution patterns with

a.

b.

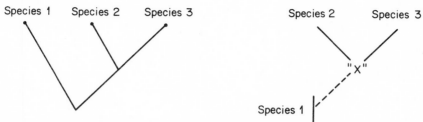

Figure 4.7 Diagram illustrating the more general nature of cladograms and phylogenetic trees from the point of view of specification of number of speciation events represented in the system. (a) A cladogram in which species 1 is the sister-group of species 2 + 3. Any number of speciation events between species 1 and species 2 + 3 may in fact have occurred and are allowed by the cladogram. (b) One phylogenetic tree derivable from the cladogram. This tree depicts the hypothesis that species 1 gave rise to an unknown descendant species ("X"), which divided to produce species 2 and 3. Thus the tree specifies exactly two actual speciation events. Pattern a (the cladogram) is thus a more general statement than pattern b (the phylogenetic tree).

respect to one another. Thus, in the Recent biota, in a region where two species are found, if these two species have been demonstrated to be more closely related to each other than either is to any other known taxon, the possibility exists that one of the species is ancestral to the other, or that both evolved from a single, immediate common ancestor. For example, in a case discussed in chapter 3, Platnick and Shadab (1976) described the spider species *Zimiromus penai* and *Z. brachet* from Ecuador, alleging their closer relationship to each other than to any other known species. The possibility is immediately raised that a speciation event connects the two species. Further, in the case of the four nominal subspecies of *Phacops rana* discussed by Eldredge (1972; see chapter 3), the pattern of synapomorphies (see figure 4.8), the absence of close relatives within the depositional basin, the relative continuity of presence of the monophyletic unit (i.e., the whole group) over a period of some 8–10 million years, and the pattern of occurrence (with respect to each other) of the constituent taxa in space and time raise the distinct possibility that they are linked by direct patterns of ancestry and descent. Indeed, Eldredge (1971) and Eldredge and Gould (1972) assumed such a tree to exist (see figure 4.9) to serve heuristically as a means of discussing patterns of speciation in geological time (see

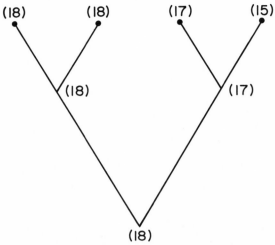

Figure 4.8 A cladogram of relationships among four subspecies of *Phacops rana*. Numbers in parentheses refer to the number of vertical columns of lenses in the eye typical of each subspecies or that hypothesized as characteristic of the common ancestor. Other patterns of synapomorphy support the cladogram. 18: *Phacops rana crassituberculata* and *P. rana milleri;* 17: *P. r. rana;* 15: *P. r. norwoodensis.*

below for further discussion of this example). All theories pertaining to speciation mechanisms hinge on field observations of populations and species which, at least implicitly, have been analyzed in precisely these terms, i.e., as ancestors and descendants. In spite of the additional assumptions required before phylogenetic trees can be constructed, phylogenetic trees appear to be crucial for further refinement of theories of the nature of the evolutionary process, specifically that part of theory devoted to the origin of species (see chapter 6; Wiley 1979). Thus these assumptions, in the long run, appear to be worthwhile, unreasonable as they might appear in any particular case.

The second assumption, that *character reversal does not occur in phylogenesis,* is vital for the evaluation of conflicting trees. Let us assume that a comparative analysis of distributions of synapomorphies has been performed. Emerging from that analysis are several putative species, plesiomorphic or equally synapomorphic in all respects to their hypothesized sister species. Such species are good

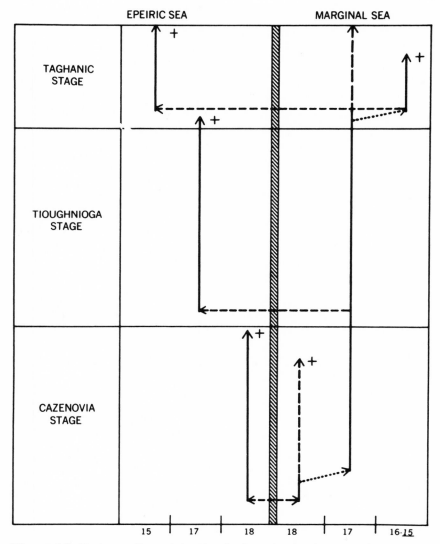

EPEIRIC SEA   MARGINAL SEA

TAGHANIC STAGE

TIOUGHNIOGA STAGE

CAZENOVIA STAGE

15   17   18   18   17   16-15

Figure 4.9 Phylogenetic tree showing the hypothesized pattern of ancestry and descent among subspecies of *Phacops rana,* derived from the clado-gram of figure 4.8 and consistent with all available extrinsic and intrinsic data. Numbers at the base of the diagram are as in figure 4.8. Dotted lines represent origin of a taxon with a new (reduced) number of columns of lenses by type b allopatric speciation; horizontal dashed lines represent migration; vertical lines represent persistence of ancestral taxon in a portion of the marginal sea other than that in which the descendant taxon occurs. Crosses denote final disappearance. (From Eldredge 1971, figure 5, and Eldredge and Gould 1972, figure 5–8.)

candidates for being ancestors to their more derived sister species. Though there are, as we shall discuss, some relatively weak tests of such hypotheses available based on extrinsic data, it is desirable to evaluate the hypothesis on the basis of intrinsic properties.

If a taxon is plesiomorphic (or at least synapomorphic) in all (analyzed) respects when compared with a series of closely related taxa (as determined by the presence of at least one synapomorphy common to them all), there is no way to reject any specific hypothesis of relationship of that species beyond specifying its status as the plesiomorphic sister-species of the remaining taxa in the series. For example, if species 1 of figure 4.7a is plesiomorphic in most respects when compared with species 2 and 3, there is the possibility that species 1 is the ancestor of the other two species (as in figure 4.7b). But there is no formal means of rejecting this hypothesis, at least with sole reference to the evaluation of intrinsic properties (see Engelmann and Wiley 1977; Platnick 1977a). Eldredge and Tattersall (1975) claimed that, at least in terms of cranial morphology, the early Pleistocene hominid *Australopithecus africanus* is so completely plesiomorphic that it is difficult to assess its relationships among Hominidae. There is certainly no way of formally rejecting the hypothesis that *A. africanus* is the ancestral species from which all subsequent species of Hominidae evolved (figure 4.10).

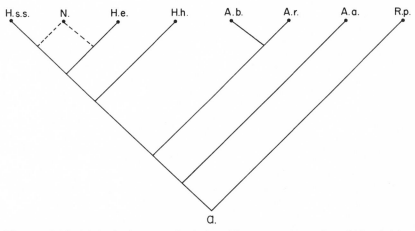

Figure 4.10 (a) A cladogram of relationships among species of Hominidae based solely on cranial characteristics. (After Eldredge and Tattersall 1975, figure 4, and Tattersall and Eldredge 1977, figure 5.)

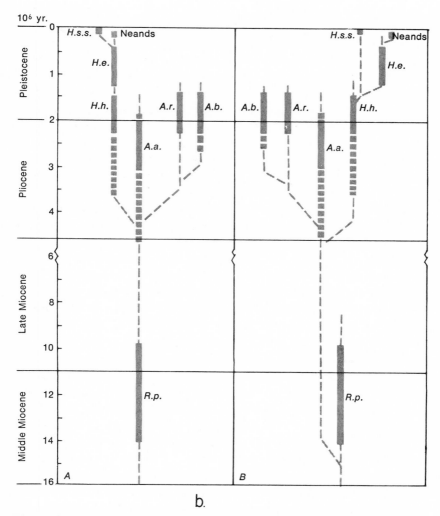

b.

Figure 4.10 (b) Two phylogenetic trees consistent with the cladogram. *R.p.*, *Ramapithecus punjabicus; A.a., Australopithecus africanus; A.r., Australopithecus robustus; A.b., Australopithecus boisei; H.h., Homo habilis; H.e., Homo erectus;* N, neandertaloids; *H.s.s., Homo sapiens sapiens. Note that,* according to the cladogram, the genus *Australopithecus* is nonmonophyletic, and alternative views of the relationships of the neandertaloids are expressed. (From Tattersall and Eldredge 1977, figure 7.)

Thus there is a profound, if formal, difficulty in the testing of hypotheses of ancestry and descent. This difficulty stems from the logical necessity of postulating that an ancestor must be plesiomorphic or equally synapomorphic with respect to its descendant—which must, by definition, be true. If a species fits these requirements, the hypothesis of ancestry is formulated. Testing of the hypothesis comes when more characters are sought. For example, demonstration that a species hypothesized as being an ancestor of another species in fact possesses an autapomorphy rejects the hypothesis of its ancestral status without rejecting the underlying cladogram. Of course, demonstration that a hypothesized ancestor shares apomorphies with a third taxon (i.e. some taxon other than the putative descendant), results in the rejection of the entire cladogram and not just the tree. Thus there is an element of testability to hypotheses of ancestry and descent among species, but this element comes at the expense of our second assumption, that *character reversal does not occur in phylogenesis* (see also the discussion of phylogenetic neoteny, chapter 2). If autapomorphies are sufficient to reject the hypothesis of ancestry, there is an implicit assumption that a particular autapomorphy cannot revert to its "primitive" condition. This is nothing more than the criterion of parsimony and, viewed in this light, perhaps not an overly steep price to pay for testability of phylogenetic trees. However, paleontologists in particular are fond of tracing character changes through vertical sequences of rocks. If a vertical sequence "documents" a character reversal, it is taken at face value. The second assumption, the invocation of parsimony in the form of a methodological rule ("character reversal does not occur"), becomes unacceptable when viewed in this context, e.g., the "stratophenetic" approach of Gingerich (1976, 1979. See figure 3 in Gingerich, 1979, for an example of character reversal within an hypothesized continuous lineage through a sequence of layered rocks). Even though plesiomorphy and apomorphy are qualities themselves hypothesized by the systematist, it is nonetheless parsimonious to assume that an apparently more apomorphous state will not, in fact, become further modified into an apparently more plesiomorphous state. Paleontologists have for years been rejecting trees on the basis that a certain taxon is "too specialized" to have given rise to another. The assumption of irreversibility is methodologically essential if hypotheses of ancestry and descent are to have any degree of testability whatever.

Cladograms to Trees: Formulation and Testing

Figure 4.11 shows the simplest form of phylogenetic tree and its most general relation to a cladogram. The case is trivial, but nonetheless epitomizes the most fundamental, formal distinction between a cladogram and a tree: the presence of taxa specified at nodes (branching points) on a branching diagram. Cladograms do not have taxa at branching points, trees do. This most general tree (figure 4.11b) might be termed an A_i tree ("A" for ancestor). Species 1 and 2 are implicitly linked on the cladogram by a pattern of synapomorphy derived from some common ancestor of unknown remoteness (figure 4.11a). In figure 4.11b, the A_i tree, the ancestor is merely made explicit. There are three possible values for i: $i = 1$ (species 1 is ancestral to species 2); $i = 2$ (species 2 is ancestral to species 1); or $i \neq$ 1,2 (neither species is ancestral to the other). If $i = 1$ or $i = 2$, there

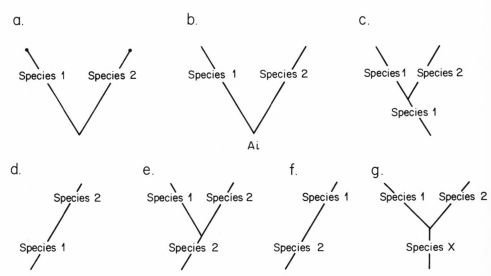

Figure 4.11 Diagram illustrating the relationship between a (a) two-taxon cladogram and (b) an A_i tree (the simplest form of a phylogenetic tree). Also shown are trees corresponding to the following values of i: (c,d) $i = 1$; (e,f) $i = 2$; (g) $i \neq 1,2$, i.e., $i =$ some other taxon (X).

are two possibilities in each case (figure 4.11c, d, e, f). If $i \neq 1,2$ (Figure 4.11g)there is a total of five potential forms of an A_i tree in any case involving two known species. The problem of converting a cladogram into a tree, then, resolves down to a matter of specifying the value of i.[3]

The process of identifying a value of i consists of considering each possible specific form of a hypothesis of ancestry and descent at each branching point (node) between sister-species on a cladogram. The simplest case involves a dichotomous node, where only two species are involved at any one branch point. Making the first assumption, that it is possible that the two species are directly related by a speciation event, the first hypothesis to be considered is that the relatively more plesiomorphous of the pair of sister-species is ancestral to the other. In other words, on the basis of the second assumption, the hypothesis that the relatively more apomorphous sister-species is the ancestor of the other is automatically tested and rejected.

In testing the hypothesis that the relatively more plesiomorphous species is ancestral to the other species, there are two possible outcomes: the hypothesis is rejected (in the general manner suggested above, i.e., by discovery of one or more autapomorphies or by rejection of the underlying cladogram itself) or the hypothesis is provisionally accepted. Should the plesiomorphous sister-species turn out to be relatively plesiomorphous or equally synapomorphous with respect to all available intrinsic data, the hypothesis that it represents the ancestor of the second species might further be tested by consideration of configurational (extrinsic) data: distributions of the two species in space and, if data are available, in time.

The simplest case occurs when the two species are living contemporaneously and are known from only a single point in time, the best example being most elements of the Recent biota. Two species known as fossils from only a very short interval of time conform equally well. In such circumstances, type *b* allopatric, parapatric, and sympatric speciation are the three major models of speciation

---

3. In this example, the distinction between notational conventions in trees is clearly seen. When species are depicted as lines, there are indeed two forms of trees for each case (i.e., for i = 1 and i = 2), a total of four possibilities (i $\neq$ 1,2 gives the fifth possibility), conforming to conflicting theories of speciation reviewed earlier in the chapter. If, however, species are represented as dots, there are only three trees possible, one each for i = 1, i = 2, and i $\neq$ 1,2.

which are theorized to produce a new species living alongside (as a contemporary of) its ancestor. Referring to figure 4.4, distributional data may be suggestive of the actual mode of speciation, and serve as a relatively weak test of the hypothesis of ancestry. All three models of speciation predict that, initially, i.e., during and right after the speciation "event" (which is hypothesized to be rapid but which may in fact take as long as several thousand years), the size of the descendant species (measured in numbers of individuals) will be small with respect to that of the ancestral species. Thus should the apomorphous sister species, hypothesized to be the descendant, have a restricted distribution with respect to its putative ancestor, and should the two species display disjunct distributions, the data are consistent with an interpretation of type *b* allopatric speciation, and the hypothesis of ancestry and descent may be considered to have been corroborated somewhat further. However, size and distribution of descendant species relative to their ancestors emerge as predictions from theory only in the early history of the descendant species. Distributions of both descendants and ancestors may or may not change subsequent to speciation. No element of speciation theory can predict subsequent patterns of distribution in detail (see chapter 6 for further discussion of the prediction of distributional attributes of species). An initially allopatric descendant may come to range as widely as, and even become fully sympatric with, its ancestor (figure 4.4). Moreover, there are no criteria whereby relative recency of speciation—which would allow more confident predictions about the significance of distributional data—can be assessed. Spatial distributional data for two contemporary species offer no firm means of testing hypotheses of ancestry and descent.

The only direct prediction that arises from the notion of ancestry and descent is the rather obvious one that the ancestor must have been older than its descendant. There must have been some segment of time during which the ancestor was in existence, and its descendant was not. Fossils offer the only direct evidence of the existence of taxa in pre-Recent times. Shaw (1964) has pointed out that the actual temporal range of any fossil species must be considered unknowable. The observed stratigraphic range, especially in any local area, cannot be assumed to be the total life-span of a species. What is sampled is actually a portion, somewhere in the middle, of the total life-span. Similarly, total geographic distributions of fossil taxa are virtually unknowable. In many instances, sediments were not even deposited in all areas where a species might have been living. This is especially true for terrestrial organisms, but pertains to aquatic organisms as well. And where deposition did occur, there is no guarantee that individuals of a given species are fossilized even if they had been living there. Moreover, erosion has destroyed at least portions of many sedimentary rock units, particularly around the margins of sedimentary basins (where speciation might be predicted to occur most frequently according to the model of type *b* allopatric speciation). Or sediments may remain buried under miles of younger rock, wholly inaccessible to the systematist. Subduction, metamorphism, and diagenesis (geochemical alteration of sediments and fossils) are additional factors that limit the credibility of temporal and spatial distributional data of fossil species. In an hypothesis as specific and detailed as those postulating derivation of one species from another, the fossil record can rarely if ever be taken literally.

Nevertheless, if a plesiomorphic sister species is always found in younger (independently dated) rocks than its hypothesized descendant, then, even though the hypothesis is not decisively rejected, there would seem to be little reason to retain the hypothesis that the younger species is ancestral to the older one. Such data, however, have no bearing on the cladogram: the younger but more plesiomorphous species remains the plesiomorphous sister of the older, more apomorphous species.

The phylogenetic tree (figure 4.9) for the *Phacops rana rana* species-group depicts two putative instances of type *b* allopatric speciation. For example, *P. rana rana* (number 17 in figures 4.8 and 4.9) is derived with respect to *P. rana crassituberculata* (number 18 on the diagrams). *P. r. crassituberculata* is known from slightly older rocks in eastern North America. *P. rana rana* is first known to occur in slightly younger rocks representing near-shore environments in central New York State. In one quarry, a few specimens have been collected which are morphologically intermediate between the two taxa. Higher in the same quarry, undoubted specimens of *P. rana rana* occur alone. In still younger sediments, *P. rana rana* is found throughout an approximately 2 million-year interval of time, in the near-shore environments preserved in the present-day Appalachian Mountains. During that same temporal interval, *P. r. crass-*

ituberculata became restricted to the more offshore waters of what is now the American Midwest. Geological evidence suggests that the seaway disappeared in the Midwest at the end of this interval. When the sea reappeared, *P. rana rana* was found to be ubiquitous and *P. r. crassituberculata* was no longer found. Thus distributional histories of fossil taxa can be sequentially mapped (figure 4.12).

Eldredge (1971) interpreted the sequence as an instance of type *b* allopatric speciation, where *P. r. crassituberculata* gave rise to *P. rana rana* as a peripheral isolate. The two taxa remained allopatric while *P. rana rana* expanded its range (along the present-day Appalachians). The distributional data of this example are consistent with this interpretation. Further tests are possible by examining additional data: expansion of range data (both temporal and spatial) would serve to reject the specific form of the hypothesis. Were *P. rana rana* to be shown to occur as early as the earliest known occurrence of *P. r. crassituberculata,* or should its inferred initial geographic range be shown to have been broader, interpretation of the data as a case of type *b* allopatric speciation would be falsified.

In short, distributional data can be interpreted as consistent with a specific hypothesis of ancestry and descent, and one of the several available models of the speciation process will usually appear to fit the data rather well. But lack of strict predictability of necessary postspeciational patterns of distribution, coupled with necessarily faulty distributional data, severely restricts the use of extrinsic data as a critical test of hypotheses of ancestry and descent—phylogenetic trees. There are no strict criteria available to allow judgment as to how much anomalous distributional data can be "allowed" before a hypothesis of ancestry and descent is disallowed. Species with disjunct (whether spatial or temporal) distributions may nonetheless have a direct ancestral-descendant relationship, and there is no way of specifying the limits in distributional data beyond which such hypotheses are to be rejected.

If the hypothesis that either of two species is ancestral to each other is rejected, a fifth possibility remains to be considered. Type *a* allopatric speciation produces two descendant species simultaneously from a single ancestor. Thus, in a situation involving two taxa, the fifth possibility, that both species were derived from an immediate common ancestor, results directly in a tree where an ances-

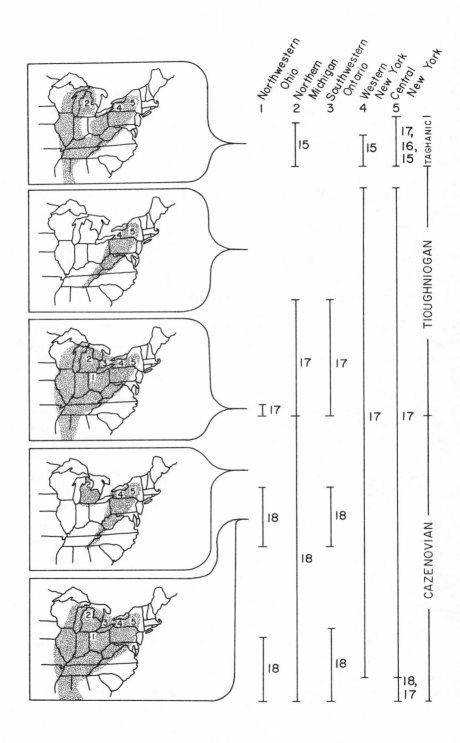

tor is postulated to have existed, but is unknown. Such a hypothesis further states that, whatever that ancestral species was, it could not have been either species 1 or species 2. Such a tree is a real tree, not a cladogram, simply because it does postulate a direct, albeit unknown, ancestor.

A three-taxon statement (a cladogram) may be investigated as a source of the identity of the unknown ancestor. There are several possibilities. In the cladogram of figure 4.13a, the possibility is considered that species 3 is ancestral to both species 1 and 2. But, according to the cladogram, species 1 and 2 share a synapomorphy not present in species 3. Thus their common ancestor must have possessed that synapomorphy, and species 3 is thereby automatically rejected as their immediate common ancestor. The only kind of cladogram involving more than two species that could resolve into a phylogenetic tree representing type *a* allopatric speciation is that in figure 4.13b. Species 3 is discovered or considered after species 1 and 2 are depicted as sharing an unknown common ancestor. There are no synapomorphies linking any two species as a subset of the three. Trichotomous cladograms may be considered unresolved dichotomous cladograms (i.e., as if the cladogram of figure 4.13b may ultimately be shown to have a structure like that of the cladogram of figure 4.13a). This position is especially pertinent to theories of relationship among three monophyletic taxa of generic or higher rank. Insofar as species are concerned, there is a further possibility: a trichotomous cladogram might represent the closest approximation to the actual evolutionary relationships among the three species. The expected outcome of type *a* allopatric speciation is precisely this form of cladogram. All that is required is that the three species be separately diagnosable, i.e., that species 1 and 2 (the putative descendants) each have at least one autapomorphy. In such a situation, the hypothesis is tested that the newly discovered or consid-

Figure 4.12 Five sequential maps showing distribution of seaways in eastern North America in the Middle Devonian Period. Numbers on the map refer to five representative localities. Vertical lines show the stratigraphic occurrence of different subspecies of *Phacops rana,* represented by numbers of columns of lenses in the eye (as in figures 4.8 and 4.9). Blank spaces indicate intervals when a sea was absent from a locality. Major subdivisions of the Middle Devonian of eastern North America are shown; a time span of approximately 10 million years is represented. (From Eldredge and Eldredge 1972:55.)

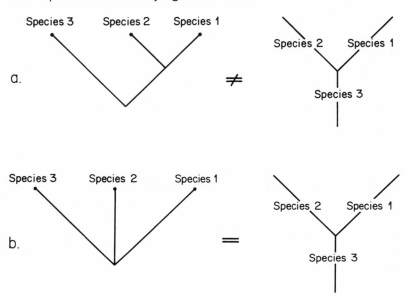

Figure 4.13 Diagram illustrating relationships between cladograms and phylogenetic trees for cases involving three species. (a) Cladogram in which species 1 and 2 are the sister-group of species 3. Such a diagram implies possession of at least one synapomorphy held by species 1 and 2, but not species 3. Since the synapomorphy must have occurred in the common ancestor, it is impossible to construct a phylogenetic tree with species 3 directly ancestral to species 1 and 2. (b) A trichotomous cladogram among three species. Such a situation could represent type *a* allopatric speciation, in which a single species (e.g., species 3) splits into descendant species 1 and 2. All that is required is that the three species be separately diagnosable and that one be plesiomorphic or equally synapomorphic with respect to each of the other two species. Species 1 and 2 do not uniquely share any synapomorphies in this instance.

ered species 3 is the ancestor of both species 1 and 2. The procedure is exactly the same as that outlined above for the hypothesis that $i = 1$ or $i = 2$. Given this possibility, further speciation could produce taxa of higher rank ultimately linked as a genuine polytomy. Thus not all polytomous cladograms need to be mere expressions of ignorance.

Actual instances of simultaneous polytomous splitting, in which two or more species are differentiated (by type *b* allopatric, parapatric, or sympatric speciation) at the same time, have the result that no two descendant species will share a synapomorphy; thus, again, the cladogram will perforce be polytomous, and the situation will again be as in figure 4.13b. The only expected difference will be in the form of the phylogenetic tree: the ancestral species will be expected to survive as a contemporary of its descendants for some unpredictable length of time. White (1968) and his co-workers have proposed just such a circumstance involving morabine grasshoppers of the *viatica* complex of Australia (see figure 4.14).

Phylogenetic trees are thus more specific and complex hypotheses than cladograms. They rely on two additional assumptions: (1) species at hand are directly related by actual speciation events and (2) character reversal does not occur. The first assumption is necessary for the analysis to proceed, and the second is necessary to confer a degree of testability to phylogenetic trees, i.e., to make them scientific hypotheses. But if the goal of systematics is the recon-

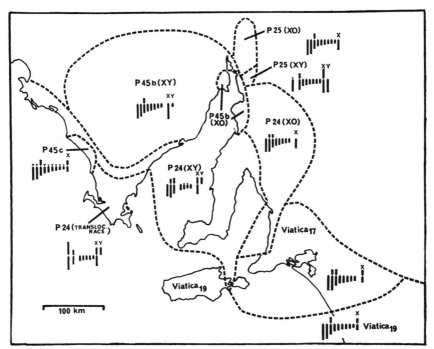

Figure 4.14 Map of a portion of southern Australia, showing distribution of certain taxa of the *viatica* group of morabine grasshoppers. Species are identified by chromosome number (e.g., P45c); haploid karyotypes are also shown. P45c appears to be the closest to the ancestral *"viatica"* type. (From White 1968, figure 1. Copyright 1968 by the American Association for the Advancement of Science.)

struction of the history of life, then specification of ancestral-descendant relationships among the phylogenetic units of evolution—species—remains at least ideally desirable. Trees express interpretations of actual patterns of descent and are thus vital to the further analysis of the nature of the evolutionary process. It is at this level that specific theories of the nature of the evolutionary process have a relationship with systematics: phylogenetic trees must be consistent with both cladograms and concepts of evolution. Cladograms, on the other hand, merely require the hypothesis that life, somehow, has evolved.

Chapter
5

Biological Classification

IN its most elemental form, classification is the grouping of objects into a larger more inclusive set; alternatively, classification can be viewed as the subdivision of a larger set into smaller subsets. In classifying there is always a reason for grouping some objects together while at the same time excluding others, and that reason must be based, ultimately, on similarity. Words, or, perhaps more strictly, the meanings of words, are themselves classifications. The process of classifying is thus a very ancient activity within human history: "From earliest times the human mind sought out the elements of order in the world, and the first step in this direction consisted in the noting of similarities between things. Such noting of similarities between things constitutes an implicit, if not an explicit, classification of them" (Wolf 1930:139).

The Greek philosophers provided the foundation of ideas from which subsequent views of classification are derived. The philosophy of essentialism, originating with Plato and Aristotle, held that a collection of objects can be defined as a set (called by them a "species") when each member of the set shared the same essence: "In Aristotle's view three things can be known about any entity—its essence, its definition, and its name. The name names the essence. The definition gives a complete and exhaustive description of the essence" (Hull 1965:6). The definition of the essence, then, was a list of jointly held properties—similarities—that justified applying a name to the set of objects. Essences were considered *real* entities in nature, not artifacts of human thought; consequently, groups characterized by such essences were *natural*. The problem of "natural classification," i.e., the discovery of "natural" groups, has captured the interest of biologists since Aristotle.

Carl von Linné constructed his now famous system of botanical classification from the standpoint of Aristotelian philosophy (see Larson 1968). His "essences" were extracted from the varieties in form of the reproductive system of plants. Soon thereafter, the French zoologist Georges Cuvier took a similar approach using "functional criteria" to identify the most important characters (essences) that could serve to delineate groups (Coleman 1964). Belief in natural classification was widespread throughout seventeenth-, eighteenth-, and early nineteenth-century biology, and most biologists of these times followed a simple rationale: natural groups are those created by God and it is the task of the pious scientist to discover and make them manifest.[1]

The observation that natural historians have always been concerned with creating "natural" classifications is important, for it suggests a commonality in purpose that transcends the intellectual dogma hegemonic at any particular period of time. In a real sense, the problem of classification, in particular, is the problem of comparative biology in general: the search for natural groups. Indeed, it has been, and always will remain, one of the fundamental questions and pursuits of biology, for the simple reason that it reflects mankind's curiosity about reconstructing the history of life.

Historically, the problem of classification—the search for natural groups—preceded evolutionary theory and thus may appear to be largely independent of it. This, of course, is not the conclusion most contemporary systematists might draw were the question put to them. But consider. The search for natural groups is a search for pattern. Darwinian evolutionary theory is a theory about process, not pattern analysis, and the question of mechanism can be shown to have a tenuous link to the methods used to discover natural groups, i.e., to reconstruct the pattern of life's history (see chapter 2 on cladistic analysis). This is not to assert, of course, that evolution has not produced the pattern—natural groups. But has it been demonstrated that the introduction of evolutionary thought altered the *method* of systematics to any great extent? Louis Agassiz, for example, constructed a "genealogical" diagram (Figure 5.1) 15 years prior to *The Origin of Species,* a diagram that does not differ materially in its

1. During the eighteenth century, French botanists sought to discover "natural order" by an essentialist criterion of group similarity: those plants most like one another in their "essential characteristics" were combined into "natural" groups (Burkhardt 1977:49). The classificatory system of Linnaeus was considered "artificial" in that it was based upon the characters of the sexual system alone.

content from those drawn after the introduction of evolution. Only the interpretation was to change.[2]

By the beginning of the nineteenth century, the search for natural groups had become more intense: "The belief in natural classification rose to dominance with remarkable swiftness and lack of debate. Perhaps the accumulated experience of taxonomists had built up until they knew that they must be on the right track. They sensed that their efforts were revealing nature herself, and by 1800 they felt this with new confidence" (Winsor 1976:3).

That nature might be hierarchical in arrangement, rather than a linear chain, began to be considered by nineteenth-century biologists. The famous Swiss botanist A. P. de Candolle, for example, stated in 1813 that the pattern of nature is characterized by groups within groups (Winsor 1976). This belief gained further acceptance prior to Darwin, and while most biologists of the period relied on Scripture to provide the ultimate causal explanation of these "patterns in nature," the actual recognition of natural groups was expressed through classification. Ultimately, it was left to Darwin to give a new meaning to the concept of natural groups.

Darwin, Natural Groups, and Classification

In chapter 13 of *The Origin of Species,* Charles Darwin presented a discussion of the principles of classification and articulated a posi-

2. Patterson (1977:580) notes of Agassiz's diagram: "Agassiz's example shows clearly that belief in evolution is not necessary for the production of such diagrams. . . . The information contained in these diagrams is therefore not necessarily concerned with evolution or phylogeny."

Several systematic botanists have also had similar perceptions about the introduction of evolutionary thought: "It seems to be too readily assumed that the doctrine of evolution is the basis on which classification builds, whereas in practice it is rather the reverse. . . . Even without any theory [of evolution], taxonomy would proceed as it always has done, with the aim of classifying organisms in the most convenient manner, which is to place together obviously related genera, species, and other groupings" (Ramsbottom, cited in Wilmott 1950:44). Wilmott (1950:44) also notes: "The fact remains that systematists had produced a rather surprisingly natural system before Darwin."

Zoologists, too, have commented on the fact that actual taxonomic practice was little affected by the introduction of evolutionary thought. *The Origin of Species,* says Hopwood (1950b:59), has had "little effect on the principles of taxonomy, which remain much as they were in the days of Linnaeus, of Lamarck, of Cuvier, and of S. P. Woodward."

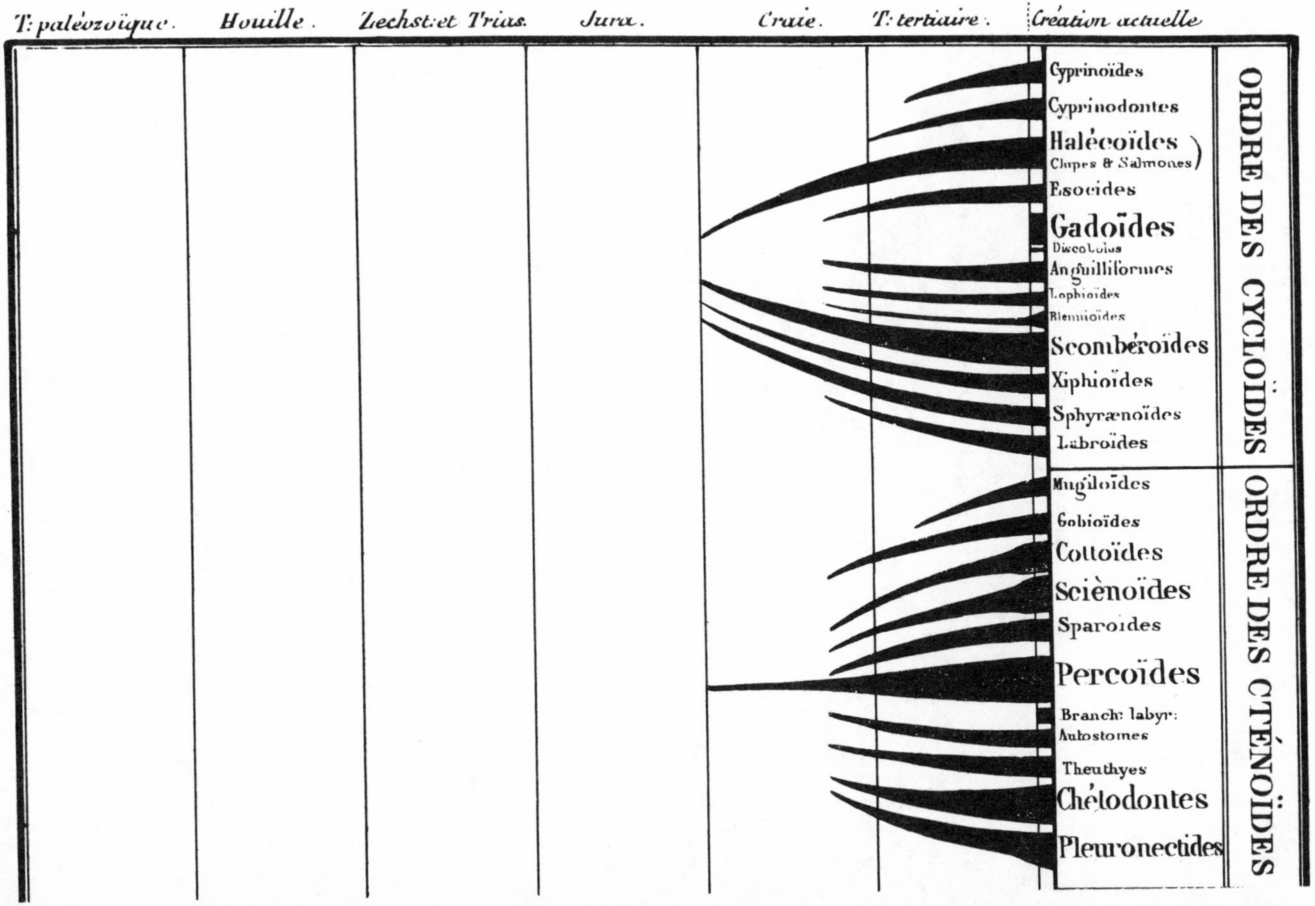
T: paléozoïque.
Houille.
Zechst: et Trias.
Jura.
Craie.
T: tertiaire.
Création actuelle
ORDRE DES CYCLOÏDES
Cyprinoïdes
Cyprinodontes
Halécoïdes
Clupes & Salmones
Esocides
Gadoïdes
Discobolus
Anguilliformes
Lophioïdes
Blennioïdes
Scombéroïdes
Xiphioïdes
Sphyrænoïdes
Labroïdes
ORDRE DES CTÉNOÏDES
Mugiloïdes
Gobioïdes
Cottoïdes
Sciènoïdes
Sparoïdes
Percoïdes
Branch: labyr:
Aulostomes
Theuthyes
Chétodontes
Pleuronectides

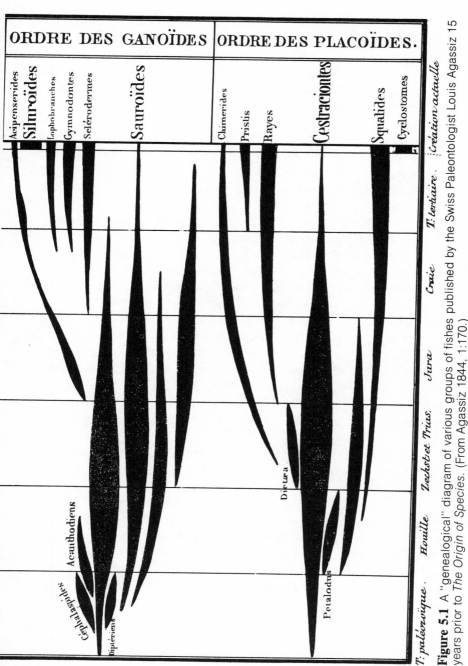

ORDRE DES GANOÏDES

Acipenserides
Siluroïdes
Lophobranches
Gymnodontes
Sclérodermes
Sauroïdes

Acanthodiens
Céphalaspides
Diptériens

ORDRE DES PLACOÏDES.

Chimerides
Pristis
Rayes
Cestraciontes
Squalides
Cyclostomes

Dictea
Petalodus

T. paléozoïque. — Houille — Zechstet — Trias. — Jura — Craie — T. tertiaire. — Création actuelle

Figure 5.1 A "genealogical" diagram of various groups of fishes published by the Swiss Paleontologist Louis Agassiz 15 years prior to *The Origin of Species*. (From Agassiz 1844, 1:170.)

tion that, despite some controversy through the years, has persisted until the present. Recently, there has been considerable debate as to exactly what Darwin meant, particularly with respect to how closely his views can be reconciled with one or the other of the systematic philosophies currently in conflict. Thus, evolutionary systematists (Simpson 1959, 1961; Mayr 1969, 1974) have interpreted Darwin's writings as indicating classificatory principles similar to their own. On the other hand, Nelson (1971a, 1974) has taken issue with this interpretation and suggests that Darwin was, in principle, closer in philosophy to the school of phylogenetic classification. It might seem unimportant, except from a historical point of view, whether Darwin's position on classification was closer to this or that modern school of thought. From the standpoint of justifying a particular modern-day philosophy of classification, such a debate does indeed seem unimportant, since modern conflicts in theory and method must be decided on merit, not historical precedent. However, a close examination of Darwin's views is important here because they have had a significant impact on subsequent classification theory, and can help us understand a fundamental dilemma that has characterized systematic practice since the time of Aristotle (this is the dichotomy of A/not-A groups, to be discussed shortly). There is one additional reason: the principles enunciated in this chapter can be seen to have their germination with Darwin, and our discussion will acknowledge this debt. We also hope that this discussion will clarify aspects of the current debate over Darwin's philosophy of classification.

Darwin perceived the pattern of organic diversity to be hierarchically arranged (1859:411): "All organic beings are found to resemble each other in descending degrees, so that they can be classed in groups under groups. This classification is evidently not arbitrary like the grouping of the stars in constellations." Referring to the only diagram in *The Origin of Species* (our figure 5.2), he makes this notion of hierarchical arrangement more explicit, and for the first time relates it to classification. Darwin's detailed interpretation is at the center of the contemporary debate and hence is quoted here in its entirety:

> I request the reader to turn to the diagram [our figure 5.2] illustrating the action, as formally explained, of these several principles [i.e., of his theory of evolution]; and he will see that the inevitable result is that

the modified descendants proceeding from one progenitor become broken up into groups subordinate to groups. In the diagram each letter on the uppermost line may represent a genus including several species; and all the genera on this line form together one class, *for all have descended from one ancient but unseen parent, and consequently, have inherited something in common.* But the three genera on the left hand have, on this same principle, much in common, and form a sub-family, distinct from that including the next two genera on the right hand, which diverged from a common parent at the fifth stage of descent. These five genera have also much, though less, in common; and they form a family distinct from that including the three genera still further to the right hand, which diverged at a still earlier period. And all these genera, descended from (A), form an order distinct from the genera descended from (I). So that we here have many species descended from a single progenitor grouped into genera; and the genera are included in, or subordinate to, sub-families, families, and orders, all united into one class. Thus *the grand fact in natural history of the subordination of group under group,* which, from its familiarity, does not always sufficiently strike us, is in my judgment fully explained. (Pp. 412–13; italics added)

Three points are worth stressing. Darwin clearly views the evolutionary process as producing a hierarchical pattern of nested sets of taxa (his "subordination of group under group"). Moreover, species are grouped into sets (genera, families, and so forth) because they all have descended from a single ancestral species. Finally, he implies, and quite strongly, that the species descended from a common ancestor are united by features inherited from that ancestor. Later in the chapter he emphasizes this last point, again referring to his illustration: "All the modified descendants from A will have inherited something in common from their common parent, as will all the descendants from I; so it will be with each subordinate branch of descendants, at each successive period" (p. 421).

From these statements it would seem Darwin was outlining a viewpoint foreshadowing the more explicit formalization of Hennig (1966): that the pattern of the phylogenetic process is one of nested sets of taxa defined in terms of nested sets of shared derived characters (synapomorphies). And, like Hennig, Darwin was quick to emphasize the distinction between grouping on the basis of a general measure of similarity and the special kind of similarity that is synapomorphy. Darwin also saw the genealogical pattern, manifested by shared derived similarity, as providing the long sought key to a

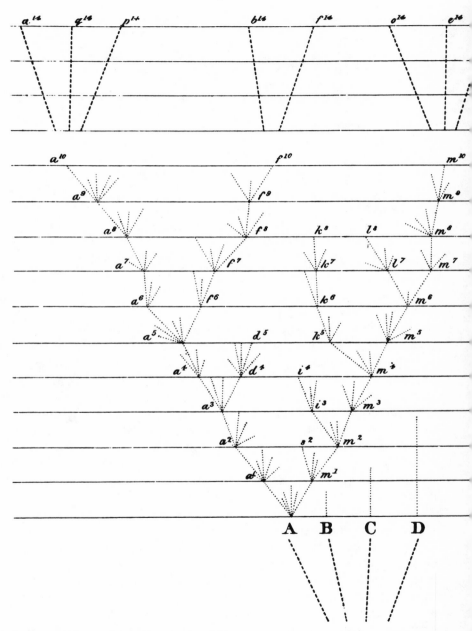

Figure 5.2 Darwin's phylogenetic diagram on which he based his discussions of

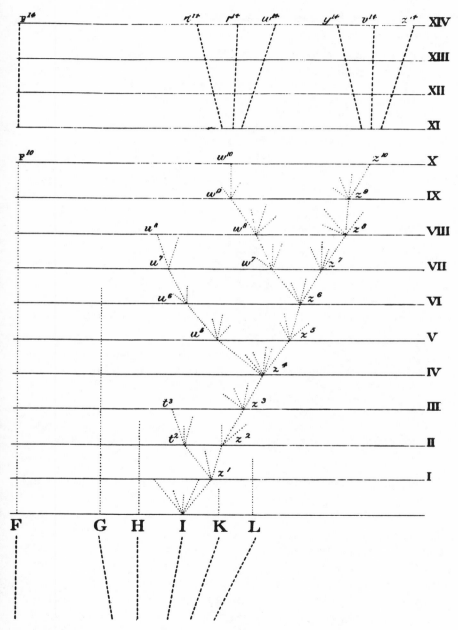

...chical structure of evolutionary history and classification. (From Darwin 1859.)

"Natural System." Indeed, Darwin was the first to propose grouping by descent rather than by general similarity (pp. 413–14):

> But what is meant by this system? Some authors look at it merely as a scheme for arranging together those living objects which are most alike, and for separating those which are most unlike. . . . But many naturalists think that something more is meant by the Natural System; they believe that it reveals the plan of the Creator. . . . Such expressions as that famous one of Linnaeus . . . that the characters do not make the genus, but that the genus gives the characters, seem to imply that something more is included in our classification than mere resemblance. I believe that something more is included; and that propinquity of descent—the only known cause of the similarity of organic beings—is the bond, hidden as it is by various degrees of modification, which is partially revealed to us by our classifications. (Pp. 413–14)

For Darwin, and many naturalists before him, the concept of natural classification meant that the perceived pattern of groups within groups must be translated into a Linnaean hierarchy as faithfully as possible. Darwin had some specific thoughts on how this was to be accomplished, and his discussion has been variously interpreted by modern systematists:

> I believe that the *arrangement* of the groups within each class, in due subordination and relation to the other groups, must be strictly genealogical in order to be natural; but that the *amount* of difference in the several branches or groups, though allied in the same degree in blood to their common progenitor, may differ greatly, being due to the different degrees of modification which they have undergone; and this is expressed by the forms being ranked under different genera, families, sections, or orders. (P. 420; italics in original)

What exactly did Darwin mean here? He referred the reader to his diagram. He supposed that A through L were related genera living in the Silurian; all were descended from a single species in the remote past. Only three genera (A, F, and I) are considered further, and the question of interest is how to classify the Recent species a^{14} to z^{14} (the superscript refers to the fourteenth time level since their origin from A, F, and I). The derivative species of ancestors A and I are supposed by Darwin to have diverged morphologically from their respective ancestral species, whereas the species of genus F^{14} have not diverged significantly. Lineage F exhibited no branching and little change. Darwin suggests that there should be recognized an order for lineage A with three families, an order for lineage I with two families, and he suggests that the Recent genus F^{14} should "rank with the parent-genus F." Although he does not state so explicitly, Darwin would presumably place F and F^{14} as separate genera in their own order. This arrangement, as Darwin notes, is genealogical, and the differences in ranking between lineages A, F, and I are minimal and reflect the frequency of branching more than morphological divergence. That Darwin argued against using morphological divergence to delimit classificatory groups was illustrated by the following example: "If it could be proved that the Hottentot had descended from the Negro, I think he would be classified under the Negro group, *however much he might differ* in colour and other important characters from negroes" (p. 424; italics added).[3]

Darwin's major efforts at classification involved barnacles, and it was with this group that he had to confront the realities of the practicing systematist. Genealogies within the group were not easy to discern, and this ambiguity found its way into his classification attempts. Ghiselin and Jaffe (1973) have shown that not all of Darwin's groups were based strictly on genealogy and that he sometimes gave greater consideration to differences between groups than to their similarities: "Darwin seems to have erected a system which compromised between evolutionary principles and utility. But he was formulating a system for his times, and the times were not yet ripe for a strictly genealogical arrangement" (Ghiselin and Jaffe 1973:138). Nevertheless, in principle, Darwin clearly advocated genealogical systems and, in view of this, it would be historically inaccurate to claim him the founding father of the school of evolutionary classification as advocated by Mayr and Simpson (see below). In his theoretical writings he certainly can be placed near the proponents of phylogenetic classification, despite the fact he was unable to achieve his theoretical ideals in practice (Ghiselin and Jaffe 1973:139).

Although Darwin unquestionably advanced the meaning of natural classification, perhaps his failure to obtain such a system in practice stems in part from retention of a conceptual tradition that has its

---

3. Darwin (1859:426) seems to emphasize this point in another, later passage: "We can understand why a species or a group of species may depart, in several of its most important characteristics, from its allies, and yet be safely classed with them. This may be safely done, and is often done, as long as a sufficient number of characters, let them be ever so unimportant, betrays the hidden bond of community of descent."

origins in Aristotelian philosophy and continues to the present. That tradition is characterized by attempts to dichotomize groups of organisms into mutually exclusive sets, *A and not-A sets.* Historically, a distinction has not generally been made between constructing sets of taxa on the one hand and classifying them on the other; the procedures were identical in the minds of most workers during their attempts to obtain a "natural" system. A major theme of this chapter is that failure to appreciate the conceptual and methodological implications of the A/not-A dichotomy is a primary reason for the slow progress that has been made in achieving natural classifications.

Classification and the Dichotomy of A and not-A Groups

"A groups" are characterized as those groups defined by the possession of shared derived character-states (synapomorphies), whereas "not-A groups" are those formed either on the basis of lack of characters defining A groups or by sharing character-states contrasting to those defining A groups. In either case, not-A groups are recognized most generally because of possession of the primitive condition. (It is not implied, of course, that all groups defined in this way are necessarily non-monophyletic, only that they generally are.)

The history of A and not-A groups is as old as classification itself. There never has been a time when efforts to classify organisms according to the A/not-A dichotomy have not been commonplace. Such a conception of natural order formed the basis for the first biological classifications and even today remains dominant—indeed, the contemporary school of evolutionary classification, as will be discussed, extolls the virtues of the A/not-A conceptualization.

The presence of A and not-A groups in biological classification originated with man's earliest attempts to develop a logical approach to thought processes. Specifically, such methods of grouping were a natural outgrowth of the "Laws of Thought" including the "Law of Contradiction" (*an object cannot both be A and not be A*) and the "Law of the Excluded Middle" (*an object must either be A or not be A*). Aristotle made the first significant attempt at a classification, and organized it in terms of A and not-A groups (Lamarck 1914; Hopwood 1950a):

Animals with blood (the A group)
Viviparous quadrupeds
Oviparous quadrupeds
Fishes
Birds

Animals without blood (the not-A group)
Mollusks
Crustaceans
Testaceans
Insects

Aristotle's A and not-A classification dominated biology until the work of John Ray (1627–1705), who extended the classification to include other organisms and increased its complexity but did not change its fundamental logical structure (Hopwood 1950a:28–29). (See the cladogram on page 160.)

The classification of Linnaeus did not alter Ray's scheme in any fundamental way with respect to the defining properties of the various groups. The class Vermes of Linnaeus, e.g., was reserved for those animals lacking both skeletons and articulated legs.

The early nineteenth century was a time when classifications began to take a truly modern form. Primarily through the efforts of the French zoologists J. B. Lamarck and G. Cuvier, major groups of organisms were delineated and defined. However, both these scientists followed earlier tradition and admitted not-A groups into their systems (see Hopwood 1950b). The most famous example, of course, is Lamarck's division of the Animalia into two major groups, the Vertebrata and Invertebrata. Lamarck (for a summary, see 1914) arranged his 16 classes of animals in an ascending series of complexity, and a class was often defined in terms of the absence of characters that defined succeeding, "higher" classes. Thus, worms lack legs, cirripeds lack eyes, mollusks lack a spinal cord, fishes and reptiles lack hair or feathers, and birds lack mammary glands.

The realization that groups which are defined in terms of "negative" characters might not be natural was seldom stated explicitly. Lamarck realized that Linnaean groups such as Vermes were "wastebaskets" and contained a heterogeneous assemblage of taxa. Nevertheless, he too continued this tradition and erected classes on the same basis. Indeed, as a general theme in classification, the rec-

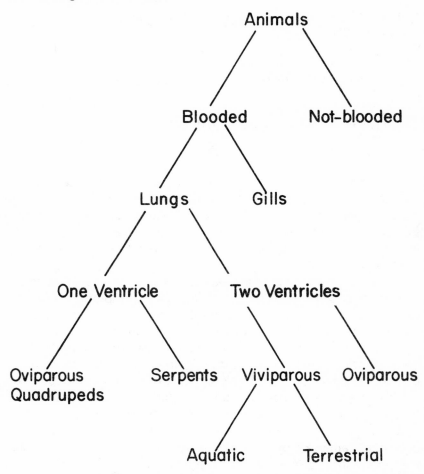

ognition of not-A groups has been accepted systematic practice, even to the present. However, there has always existed an awareness that classificatory groups should not be founded on negative characters, i.e., that not-A groups are somehow not "natural." Darwin was explicit in advocating that groups be defined by inherited similarities, which would, in effect, make those groups A groups. Although there has not always been a conscious recognition of the A/not-A dichotomy, the history of classification seems to describe a tendency to eliminate not-A groups; intuitively, biologists have recognized that not-A groups are not natural in the same way that A groups are. Nevertheless, not-A groups are still prevalent. In order to better show just how pervasive the influence of the A/not-A dichotomy has been, several examples can be considered.

Example 1. The Kingdoms of Organisms

Since ancient times, organisms have been divided into animals (sensible, motile) and plants (insensible, nonmotile)—a perfect example of A and not-A groups. This division has continued essentially without change until recent decades when technical advances permitted more detailed analyses of morphology, biochemistry, and genetics. Now, considerable discussion is ensuing as to what constitutes a "natural" classification of living organisms (see review in Margulis 1974). Despite considerable attention having been directed toward eliminating some not-A groups (e.g., placement of all "plants" together, or recognition of the dichotomy within animals of protozoans and metazoans), it would seem that current attempts at "natural" systems continue to admit not-A groups.

Most modern specialists, for example, accept a dichotomy between organisms with prokaryotic cells and those with eukaryotic cells. The prokaryotes are interpretable as a classic not-A group, as can be seen when comparing their defining properties with those of eukaryotes (from Margulis 1974, table 3):

Prokaryotes	*Eukaryotes*
Small cells	Large cells
Nucleoid not membrane-bound	Nucleus membrane-bound
No centrioles, mitotic spindle, or microtubules	Centrioles generally present, mitotic spindles, microtubules present
Sexual system absent	Sexual system generally present
No mitochondria	Mitochondria
No intracellular movement	Intracellular movement

In general, then, prokaryotes (e.g., bacteria) are defined in terms of negative characters, and this raises the question of whether they should be considered a natural group on that basis. The implication of the above definition is that they are a group of organisms, none of which can be classified as a eukaryote.

The second major division of organisms, the superkingdom Eukaryota, contains "protists" (protozoans, some algae, flagellate fungi), amastigote fungi, plants (bryophytes, tracheophytes), and animals (multicellular animals). Here, too, the concept of A and not-A groups pervades the classification, for the protists include "eukaryotes, primarily microorganisms, which are neither metazoan animals, embryophyte green plants, nor amastigote fungi" (Margulis 1974:56).

Within the kingdom Animalia, subdivisions are also organized in terms of not-A and A groups. For example, the Parazoa versus Eumetazoa; Radiata versus Bilateria; the Acoelomata versus Pseudocoelomata, versus Coelomata; and possibly even the Protostomia versus Deuterostomia. It remains to be seen whether the not-A groups of these dichotomies are natural groups in a genealogical sense.

Example 2. Classification of Invertebrate Animals

The "Invertebrata," as noted above, is a classic example of a not-A group, but at the time of its introduction near the beginning of the nineteenth century, it represented an advance over previous classificatory systems. Linnaeus, for example, recognized only insects and worms as the major groups of invertebrates, an arrangement scarcely different from that of Aristotle. But by the end of the eighteenth and beginning of the nineteenth centuries, there was a determination on the part of zoologists to produce a natural classification of these animals. In these early attempts at classification, not-A groups were prevalent, and the history of invertebrate classification through the first half of the nineteenth century appears to document an unconscious attempt on the part of systematists to eliminate such groups (see especially Winsor 1976, from which this account is largely taken).

In 1812 Cuvier recognized four major divisions, or *embranchements,* of the animal kingdom: Vertebrata, Mollusca, Articulata, and Radiata. The first three contained bilaterally symmetrical organisms and were considered easily definable in terms of the construction of their nervous and circulatory systems. The fourth group, the Radiata, was not so easily distinguishable and was characterized by the absence of a heart, a brain, and a well-developed nervous system. Furthermore, they were "rayed animals" with their internal organs displayed in a radially symmetrical pattern. The Radiata included groups as disparate as sipunculid worms, echinoderms, polyps, medusoid forms, and even "infusorians" (protozoans, rotifers). The taxonomic history of the Radiata seems a microcosm of the aspirations of systematics in general:

Cuvier himself often described the purpose of a natural classification, and its advantage over an artificial arrangement, in terms of being able to make positive general statements that would be true for all members of the group. This would not be the case for a group consisting merely of remainders, animals left over after the formation of another group, having only negative characters in common. It seems clear that the Radiata was such a collection of left-overs, and much of the work done on its classification in subsequent years can be viewed as a search for positive characters with which to define true radiates, while animals assigned that place by default were being removed to their proper home. (Winsor 1976:15).

The Radiata of Cuvier was viewed by most subsequent workers as a simplified collection of basically dissimilar forms. Lamarck (1914), for example, split the Radiata into at least four different invertebrate classes: infusorians, polyps (rotifers, hydrozoans, anthozoans, sponges), radiarians (various coelenterates, echinoderms), and worms (of various kinds). Many of the subdivisions within these four classes were based on negative characters and thus constituted not-A groups (Hopwood, 1950b:54–55).

During the nineteenth century, many biologists turned to the problem of describing and comparing the vast numbers of invertebrates being discovered, and eventually a more nearly natural classification came into existence. Of particular importance were the tedious analyses of life histories of various coelenterate groups, and the discovery of alteration of generations united the polyps on the one hand and acaleph medusoids on the other. Likewise, life history and detailed anatomical investigations began to clarify the position of the echinoderms relative to other "radiate" groups such as the coelenterates. Soon it was realized that echinoderms were distinct and could be defined by their own positive characters, including their internal organization, separate gastrointestinal track, and peculiar water-vascular system. As Winsor (1976), Hopwood (1950b), and others have noted, all these advances led to the replacement of a taxonomic arrangement of large, ill-defined groups (not-A groups) by one emphasizing smaller, more "naturally" defined taxa (A groups).

Example 3. Classification of the Vertebrata

Unlike the concept of the "Invertebrata," few systematists have questioned the monophyly and hence the "naturalness" of the Vertebrata. Most of the major groups of vertebrates—mammals, birds, reptiles, fish—have their origins as taxonomic concepts back in antiquity, and their naturalness has seldom been disputed, even in modern times. But it has become increasingly apparent that much of the generally accepted vertebrate classification is in need of re-examination, and it can be suggested that in large part this has been necessitated because of a historical reliance on the A/not-A tradition. This tradition can be illustrated by the following classification, abstracted from modern general zoology, comparative anatomy, and vertebrate paleontology texts (see particularly Kent, 1965; Romer, 1962, 1966; Storer and Usinger, 1965):

    A. Invertebrata (non-Chordata)
  AA. Chordata (notochord, dorsal nerve cord, etc.)
     B. Acrania (absence of cranium, vertebrae, brain)
   BB. Craniata (= Vertebrata) (cranium, vertebrae, brain)
      C. Agnatha (absence of jaws and paired appendages)
    CC. Gnathostoma (jaws, paired appendages)
       D. Pisces (nontetrapod, "aquatic" gnathostomes)
     DD. Tetrapoda (tetrapod, "terrestrial" gnathostomes)
        E. Amphibia (anamniote egg)
      EE. Amniota (amniote egg)
         F. Reptilia (absence of advanced avian or mammalian characters)
          G. Anapsida (no temporal opening)
        GG. Other reptiles (temporal opening)
       FF. Aves
      FFF. Mammalia

A classification closely approximating this form has dominated vertebrate biology for the last half-century, if not longer. The pervasive influence of the A/not-A dichotomy is readily apparent. In recent years, however, systematists have come to question the naturalness of the not-A groups of this classification, and alternative groupings are being proposed. It is apparent, for example, that the Acrania, Agnatha, Pisces, perhaps the Amphibia, and certainly the Reptilia are groups of unrelated taxa combined on the basis of negative characters (a general review of this evidence can be found in Moy-Thomas and Miles 1971; Greenwood, Miles, and Patterson 1973; and Løvtrup, 1977). Clearly, the history of vertebrate classification shows a tendency toward the elimination of not-A sets, and the same historical trends are evident even within the vertebrates [for example, teleost fishes (Patterson 1977) and mammals (McKenna 1975)].

Plan of the Chapter

The remainder of this chapter will be devoted to a detailed discussion of the principles of biological classification, current controversies within classification theory, as reflected by differences in the procedures adopted by alternative systematic "schools," and a consideration of methods of classification. The approach of the chapter is therefore tripartite. In the first part, the logical structure of the Linnaean hierarchy itself will be considered, with an emphasis on the limitations placed on biological classification by that structure. The second part will present a general discussion of branching diagrams and their relevance for classification. Moreover, here will be included a critical analysis of several contemporary systematic schools. Finally, in the third part, the various procedures needed to produce phylogenetic classifications will be outlined and examined. The principles discussed there should provide the reader with a basis for investigating the systematic literature in more detail and producing satisfactory classifications of his or her own.

Throughout the chapter, in the expositions and critiques, runs a common theme: that biological classification, to be natural, should be based on the concept of recognizing A groups (monophyletic taxa, as defined in the chapter on cladogram analysis) and eliminating not-A groups (nonmonophyletic taxa). It is recommended that this be accomplished by classification based on the precise representation of the nested sets of taxa uncovered by cladogram analysis. Since it has been amply demonstrated in previous chapters that cladograms are more general concepts than are alternative kinds of branching diagrams, the position is adopted here that classifications based on cladograms will result in a general reference system that not only is natural (as conceived from a historical and Darwinian tradition) and of great practical utility, but also one that reflects the underlying scientific (hypothetico-deductive) structure of systematics in general.

Structure of the Linnaean Hierarchy

As a general concept, classification is a way of ordering or arranging knowledge. Biological classification is an attempt to order or arrange knowledge about biological phenomena. Such a statement leads one to ask: What kinds of phenomena are to be classified? Furthermore, what are to be the principles or methods to be adopted when producing a classification and how is that classification to be expressed?

Within biology, the last question has had a clear, almost universally accepted answer for over 200 years: biological classifications are expressed in terms of the Linnaean hierarchy. The fact that Linnaean classification has gained such wide acceptance is surprising in itself, considering the remarkably diverse systematic philosophies that have existed and continue to do so. Because of this, one might be tempted to wonder whether the structure of Linnaean classification reveals an underlying unity within systematic biology, regardless of the methods and philosophical leanings of the individual systematist. At the same time, the philosophical diversity seen within systematic biology seems to suggest that many systematists may have lost sight of the logical and biological implications of the Linnaean system itself, and consequently have not perceived the conflicts that exist between some of the proposed kinds of knowledge that might be expressed in classification and the capability of the Linnaean system to organize and express that knowledge. Actually, both observations seem to be true, and together they may explain the general allegiance of the systematic community to the Linnaean system and the incredible diversity of opinion—and controversy—concerning the purposes, methods, and uses of classification.

What is it about the Linnaean system that seems to unify such diverse approaches to systematics? What kind of basic knowledge is being expressed within Linnaean classifications? To answer these two fundamental questions, the structure of the Linnaean system itself must be considered.

Through the ages, biologists have perceived a *hierarchy of taxa,* and the Linnaean classificatory system came into being as a way to communicate that taxic hierarchy by means of an *ordering of categories.* The Linnaean system was created, not as an attempt to order all kinds of knowledge about organisms, but rather to order the kinds of organisms themselves. Consequently, Linnaean classifications are *lists of taxa.* The order of these taxa is specified by a series of categories arranged hierarchically: certain categories are included within (or are subordinate to) others, so that each category assumes a different hierarchical level, or *rank*. Linnaeus himself recognized only five categories:

KINGDOM
  CLASS
    ORDER
      GENUS
        SPECIES

Through the years, many additional categories have come into use. Simpson (1961:17), for example, lists 21 levels, commonly applied in classifications:

KINGDOM
  PHYLUM
    SUBPHYLUM
      SUPERCLASS
        CLASS
          SUBCLASS
            INFRACLASS
          COHORT
            SUPERORDER
          ORDER
            SUBORDER
              INFRAORDER
            SUPERFAMILY
          FAMILY
            SUBFAMILY
          TRIBE
            SUBTRIBE
          GENUS
            SUBGENUS
            SPECIES
              SUBSPECIES

Not all of these categories are necessarily used in any given classification, but then again, all these and more have been adopted when some systematists have attempted to classify complex groups. As a matter of practicality, the number of categories to be employed will depend upon the manner in which the systematist wishes to represent the hierarchy of taxa. This point will be dealt with in considerable detail as this chapter progresses.

Earlier chapters on cladograms, species, and trees showed how taxa can be defined and how knowledge concerning taxa can be represented in a hierarchical pattern. Taxa, at all hierarchical levels, consist of one or more species, and, as we have emphasized, species are ontologically "real" units in nature. Categories, on the other hand, cannot be said to "be real" in the same sense as taxa. Categories are conceptual-linguistic repositories of named taxa. Furthermore, categories are independent of the included taxa in the sense that named taxa may be assigned to any category and do not inherently belong in one as compared to another (this is not to say that an arbitrary placement of taxa within categories would necessarily make biological or heuristic sense). One of the keys to understanding many of the current conflicts among alternative methods of classification is to recognize this dichotomy between reconstructing the hierarchy of taxa and then using that hierarchy as the basis of classification. Systematic methodology, viewed as a whole, must proceed in one direction, from discovering the taxic hierarchy to classifying it. By classification, one simply means the assignment of the various taxa to the different categories (i.e., assignment of rank). By definition, then, taxa logically precede categories.

To a biologist with some systematic background, the above may sound trivial and self-evident, but it can be argued that such is not the case if one views a classification as a repository of knowledge about a group of organisms. If classification is a repository or a representation of our knowledge, it can be so only in the sense that it represents a hierarchy of taxa. *A Linnaean classification is nothing more than a system of names hierarchically arranged.*

There is, of course, ample historical precedent within the systematic literature for the view that classifications are devices to store and retrieve "information" about organisms:

> A classification is a communication system, and the best one is that which combines greatest information content with greatest ease of information retrieval. (Mayr 1969:98)

> In practice, a classification is a descriptive arrangement not only for conversing about its included objects, but also for storing and retrieving information concerning them. (Ross 1974:258)

If the elements of Linnaean classifications are lists of names, then the "information content" of that classification must be the direct consequence of the structure (i.e., the topology) of the classificatory hierarchy itself. What aspects of the hierarchy provide the basis for conveying information? Clearly, that information can only reside in the logical structure of hierarchical systems in general.

Such systems consist of sets within sets (see Hull 1964, and Buck and Hull, 1966, for an extended discussion of the logical structure of the Linnaean hierarchy). Information is a result of objects being classified together into sets. The hierarchical system conveys a pattern of nested sets, defined first by the *subordination* of subsets at a level or rank within a higher, or more inclusive, level (or rank), and second by the *sequencing* of sets at the same level or rank. Hence, the following hierarchical classification conveys information by subordination (the placing of orders A and B, within Class 1, families a and b within order A, and so forth) and by sequencing (families c, d, and e within order B, for example).

CLASS 1
  ORDER A
    FAMILY a
    FAMILY b
  ORDER B
    FAMILY c
    FAMILY d
    FAMILY e
CLASS 2
  ORDER C
    FAMILY f
    FAMILY g

At this point of the discussion it does not matter what the basis was for forming the above classification. What is important is that the information content of the classification is manifested by its hierarchical topology. Indeed, that topology, and thus the information, can be translated into a branching diagram (see p. 170).

What is the information contained in these "classifications," one a Linnaean hierarchy, the other a branching diagram? Quite simply, it is that the taxa (or objects, if the classification is nonbiological) form nested sets and the elements of these sets must share some set-defining properties. It is not the properties themselves that are specified by the classification, only the sets. In the above classification, five sets are specified: classes $1 + 2$, orders $A + B$ and families $a + b$, $c + d + e$, and $f + g$. This classification implies that families a and b share some property distinguishing them from c,d, and e, and that families a,b,c,d, and e share some defining property not shared with f and g. What are these properties? The classification cannot tell us.

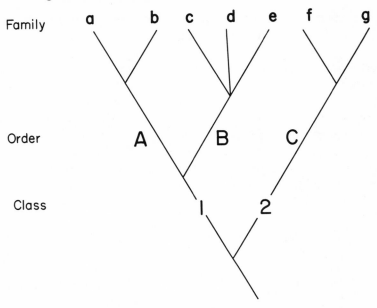

Thus *the information content of a classification is only the nested sets; nothing more can be retrieved.*[4]

The foregoing statements have two possible implications for resolving some of the arguments about methods of classification so prevalent in the systematic literature. First, it would seem that information or knowledge about organisms can be used for classification only if it can be expressed in terms of a branching diagram of the taxa under study. In other words, such information must have a logical structure equivalent to that of the Linnaean hierarchy. This may not be an important restriction, since it should be possible to represent shared properties of taxa in a hierarchical arrangement regardless of the properties or the ways in which similarities among taxa are evaluated. Second, the question of what kinds of classifications are useful within biology perhaps should be restated: what kinds of

4. Farris (1977:841) also raises this point in his response to pheneticists' arguments that phylogenetic classifications do not express any phenetic information: "No classification, even a phenetic one, is by itself any more than a collection of groups of organisms. No grouping of organisms, however arrived at, can directly express anything more about the organisms than the membership of the groups. Classifications may possess 'information content' only in the indirect sense that a classification may be used as a reference system." (See also arguments of evolutionary systematists, e.g., Johnson, 1970:233.) It can be added, in the context of the present discussion, that "reference system" can only refer to the property of set-membership.

branching diagrams are to be used as the basis for classification? As will be discussed below, there are several kinds of branching diagrams, each of which is translatable into a classification. The relevant question would seem to be: are some kinds of branching diagrams more useful than others?

Given these two points, it can be concluded that biologists who adhere to the Linnaean system in biological classification must be careful to respect the logical structure of that system. Branching diagrams representing knowledge about taxa become the focus of controversies in classification, rather than the classifications themselves. Hence, to a considerable extent, the scientific analysis of the purpose, expression, and formulation of those branching diagrams is the key issue of systematics.

Information Content and Branching Diagrams

The preceding section stated, but did not pursue in detail, two seemingly important principles of classification. First, Linnaean classifications *only* express a hierarchical arrangement of taxa. They cannot directly convey anything about the properties of taxa, or about their similarities and differences. If such information is expressed in classification, it is implied only through the mind of an individual systematist and may or may not be shared with others contemplating the same classification. Strictly speaking, then, *the information content of a classification is a hierarchy of taxa, nothing more.* As will be discussed, however, there may be some general agreement (or disagreement, even) among biologists as to the biological implications of that hierarchy.

Second, in order for Linnaean hierarchies to convey information precisely, there must be a one-to-one correspondence—an isometry—between the branching diagram of the taxa that was used as a basis for the classification and the classification itself. In other words, if classifications are to convey information—as most biologists agree they should—then common sense dictates that a classification should represent precisely the information (i.e., the taxic hierarchy) on which it is based. This one-to-one correspondence provides a basis for comparative evaluation of alternative classifications: that classification which most nearly represents the original hierarchy of taxa is best. Naturally, such an evaluation depends in part on the amount of agreement regarding the taxic hierarchy, and this in turn is amenable to scientific analysis, as discussed in chapter 2.

Although it may seem persuasive to most readers that Linnaean classifications can express information only about the structure of the taxic hierarchy, many systematists have been explicit in their belief that classifications can convey more than just this topological information. It has been stated repeatedly that classifications variously express phylogenetic relationships (genealogy), phenetic (overall) similarity, genetic similarity, evolutionary history, and so forth. Thus, to some systematists, the topological structure of the taxic hierarchy is a direct translation of statements about phylogenetic relationships or phenetic similarity. To others, classifications are to be viewed as "heuristic" systems and as schemes that can express an assemblage of evolutionary information, albeit only imprecisely.

What is to be said of these viewpoints? What kinds of information can be expressed by (or contained in) Linnaean classifications? Should not that information be readily accessible to all interested biologists, even those lacking special knowledge of the taxa in question? One cannot help but agree with Bock (1977:861): "Judgment of the usefulness of classifications is not in terms of the systematists who construct them, but with respect to the other biologists who are dependent upon classifications as the foundation for their comparative studies and the formation of their generalizations."

Consider the following hypothetical Linnaean classification:

FAMILY A
  SUBFAMILY a
      SPECIES 1
  SUBFAMILY a'
   TRIBE b
      SPECIES 2
   TRIBE b'
    GENUS c
     SUBGENUS d
     SPECIES 3
     SPECIES 4
     SUBGENUS d'
     SPECIES 5

GENUS c'
    SPECIES 6
    SPECIES 7

If there are generalizations or principles about information storage and retrieval applicable from one classification to the next, it should be possible to identify and discuss them relative to the above hypothetical classification. If, on the other hand, there are no such generalizations or principles, then it would seem impossible that any special classification of real taxa could be rationally defended in terms of its ability to store and retrieve information.

What information is expressed by this classification? Clearly, the classification is expressing something about the set-relations of seven species, and these relations are defined by the hierarchical arrangement of the categories in which those species are placed. The classification can be translated directly into a taxic hierarchy or branching diagram:

Species 1   Species 2   Species 3   Species 4   Species 5   Species 6   Species 7

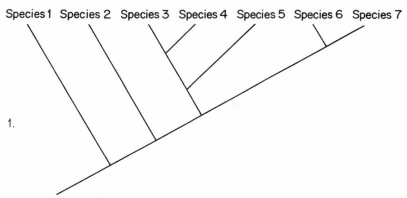

1.

But why this specific branching diagram? What is inherent in our conception of the Linnaean hierarchy that might prevent us from considering the following diagram (p. 174) or any one of the many alternatives that could be constructed?

The answer, of course, is that the information content of hierarchical arrangements, whether in the form of Linnaean classifications or branching diagrams, is expressed in terms of nested sets. Branching diagram 1 is a direct translation of the nested sets implied by the classification; branching diagram 2 is not. As a simple example, diagram 2 implies a nested set of species 5, 6, and 7, whereas the classification indicates that species 5 is nested instead

Species 1 Species 2 Species 3 Species 4 Species 5 Species 6 Species 7

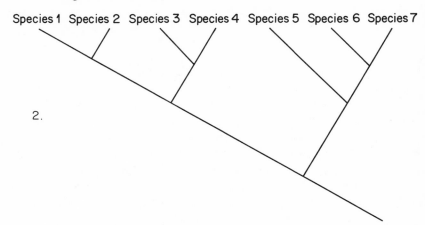

2.

with species 3 and 4. So it is evident that a Linnaean classification does contain information, in the sense that a topography of nested sets can be retrieved. What other kinds of information might be contained within that classification? The inference might be made that those taxa contained within the same set share some similarity not shared with those taxa excluded from the set. Thus, it might be supposed that species 5 shares some attribute or attributes with species 3 and 4 not shared with species 6 or 7. Consequently, some may say that the hierarchy implies the sharing of set-defining properties. If this is the case, the classification cannot tell us how many shared similarities are involved, what kinds of similarities they might be, or even that species contained within a set are actually more similar to one another than to a species excluded from the set. The only conclusion implied by the hierarchical arrangement is the possible presence of set-defining properties. However, all of the foregoing rests on the assumption, adopted prior to the investigation itself, that the hierarchy is to be characterized by set-defining properties possessed by the taxa. On the other hand, hierarchies, including perhaps the classification above, can be formed without the assumption of set-defining properties, and indeed, one branch of systematics (numerical taxonomy) advocates just that. So there would seem to be nothing inherent in an hierarchical classification (branching diagram), other than the arrangement of nested sets itself, conveying information of a special kind about the included taxa.

In the above example, a Linnaean classification was used to

derive two alternative branching diagrams. In actual taxonomic practice, of course, some conception of a branching diagram will precede the classification. The example is also illustrative of the second principle of classification referred to earlier: the information content of a classification is conveyed precisely only when there is a one-to-one correspondence with a branching diagram. If branching diagram 1 were used as the basis for constructing the classification, then the classification itself could be used subsequently by any biologist—however unfamiliar he or she might be with the group—to reconstruct the branching diagram. Obviously, the classification would not yield branching diagram 2.

Branching Diagrams and Their Role in Classification

The myriad of controversies, which for years have festered within the literature of biological classification, can be attributed, for the most part, to a single, critical issue, i.e., the kind of branching diagram that should be used as the basis for classification. Naturally, whatever branching diagram might be chosen by a systematist will convey that individual's conception about the purpose of classification.

Most modern systematists have advocated one of two systems of branching diagrams: phenograms or evolutionary trees. It will be a major thesis of this chapter, however, that a third type of branching diagram—the cladogram, as defined and discussed in previous chapters—constitutes a more general class of branching diagram and is, consequently, preferable as a basis for classification. Before pursuing that line of argument, the salient characteristics of phenograms and evolutionary trees, and their implications for classification, should be discussed.

Phenograms

Phenograms are branching diagrams depicting the *phenetic relationships* of the included taxa. By "phenetic relationships" is meant the degree of *phenetic similarity,* a measure of "overall similarity, based on all available characters without any weighting" (Cain and

Harrison 1960:3). Phenograms are the creation of numerical taxonomy (Sokal and Sneath 1963; Sneath and Sokal 1973) and traditionally are defended by numerical taxonomists as providing classifications that are "general," "natural," and preferable scientifically.
Numerical taxonomy expanded rapidly in the early 1960s, recruiting
many young practitioners, but since the end of the decade, and certainly through the 1970s, its influence in biological classification has
waned.

To summarize their methods briefly: In order to produce phenograms a large suite of characters is chosen, and the alternative
character-states of each character are coded for each taxonomic unit
(data may be qualitative or quantitative). There follow two computer
operations manipulating these basic data: first, the estimation of similarity among taxonomic units by calculating a similarity coefficient
using one of many algorithms designed for this purpose; and second, the clustering of taxonomic units into a hierarchical arrangement (a phenogram; see figure 5.3) by application of a clustering
algorithm to the similarity matrix produced by the first computer
operation. [Sneath and Sokal (1973:114–308) discuss in considerable
detail those algorithms commonly used to produce similarity matrices and phenograms; Farris (1977) presents a concise review of
phenetic principles and techniques.]

How successful are phenograms in affording a basis for classification? To judge from their own statements, numerical taxonomists have been less than satisfied with their results:

> A number of guides have been proposed for evaluating and determin-
> ing what is the "best" phenetic classification, given several potential
> candidates. However, it is clear that no complete hard and fast set of
> rules for finding the best possible classification is possible at present
> (and such a presumably ideal approach will probably not be available
> for some time to come). (Schnell 1970:294)

The reason for this pessimistic outlook, and it has been common
knowledge within systematics for some time, is that minor alterations
in numerical taxonomic procedures may radically alter the hierarchical structure of the included taxonomic units (see Minkoff 1965;
Schnell 1970, and other citations in Johnson 1970:223). Changes in
the number of taxa being studied, the choice of characters being
analyzed, the type of similarity coefficient being computed (Sokal
and Camin 1965), and the clustering technique that is adopted, all

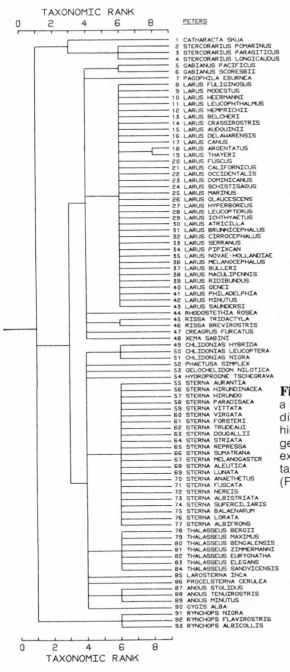

Figure 5.3 An example of a phenogram, a branching diagram showing the hierarchical pattern of general overall similarity exhibited among some taxa of birds (see text). (From Schnell 1970:36.)

have an influence on the nested pattern of taxa, that is, on the resultant phenogram. As a consequence, and seemingly as a way to elude this problem, numerical taxonomists often suggest the use of many different kinds of classifications, each to be employed for their own special purpose. But even this proposal appears questionable to some pheneticists: "While numerical taxonomy can supply more refined special classifications it is doubtful if the effort of creating them would be worthwhile in most cases" (Ehrlich and Ehrlich 1967:316).

The guides, or criteria, for choosing among alternative phenograms are mathematical in nature (e.g., see Schnell 1970:294). Sneath and Sokal (1973:278), for instance, recommend accepting that phenogram which is the "best fit" to the original similarity matrix. To measure that fit, they use a cophenetic correlation coefficient (Sokal and Rohlf 1962), which is a measure of the relative closeness among taxa implied by a given phenetic classification with respect to the actual values of overall similarity calculated for the attributes of the taxa. While such a measure may serve as an adequate procedure for evaluating phenograms, its adoption creates problems for accepting phenograms as a general classificatory scheme because, as Farris has recently shown, cladograms consistently produce higher cophenetic correlation coefficients than do the corresponding phenograms for the same set of taxa (1977:838):

> In every case the classification constructed by clustering according to special similarity [i.e., synapomorphous similarity] was found to be superior to that produced through clustering by overall similarity. The mean value of the squared cophenetic correlation coefficient for classifications produced by clustering according to overall similarity was 0.63, and the corresponding mean for analyses based on special similarity was considerably higher at 0.89 . . . *clustering by overall similarity does not seem to be an optimal classificatory method according to the principles of phenetics.* (Italics added)

These mathematical approaches to evaluating phenograms also overlook an important point: "science is *not* the process of measuring natural phenomena; that is merely a technique of science" (Johnson 1970:229; italics in original). Phenograms summarize and represent *measurements* of similarity among taxa and, as such seem to differ fundamentally from other branching diagrams (trees, cladograms) in that the latter do not manifest measurements but rather constitute hypothetical constructs capable of testing. Phenograms, trees, and cladograms all depict a nested pattern of taxa. With phenograms, the nested pattern is generated by a measurement of similarity, but not directly by a search for a nested pattern of similarity, since the different synapomorphies are not being partitioned hierarchically. With cladograms, and to a certain extent with trees, the nested sets of taxa are a direct consequence of hypothesizing a pattern about observed similarities (synapomorphies). And that pattern can be scientifically analyzed or tested: alternative patterns are evaluated with respect to the same observed similarities and "best" patterns are those with fewest conflicts in forming nested sets of taxa (see chapter 2). Choice among phenograms derives not so much from differences in the capability of those phenograms to organize pattern within a given collection of observed similarities, but rather from a calculated measure of how well a particular clustering algorithm in principle reflects the underlying structure of the similarity matrix of the taxa (itself dependent upon method). Differences in computational· methodology in effect account for variability in phenograms and, ultimately, in preferring a single phenogram as "best." In phylogenetic (cladistic) analysis, on the other hand, the same method can produce competing cladograms which are evaluated by a criterion (parsimony) not itself a consequence of the method but of hypothetico-deductive science in general. In contrast, in numerical taxonomy different methods of analysis produce alternative phenograms which are evaluated by a calculated value (the cophenetic correlation coefficient) that is linked closely to the methods themselves.

Trees

Trees comprise another group of branching diagrams, typically specifying, by one form or another, phylogenetic relationships among a set of taxa. Through the years different kinds of trees have been proposed; unfortunately, workers have seldom indicated exactly how their trees should be interpreted, hence confusion has been prevalent.

Trees commonly are constructed using three topographical features: dots, lines, and branch points (figure 5.4). Dots, when used, invariably represent identified taxa. Lines, on the other hand, may or

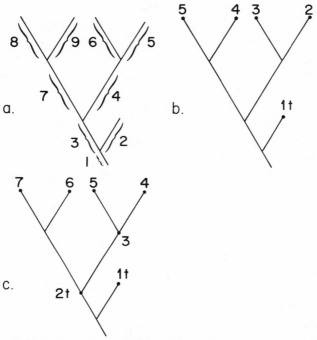

Figure 5.4 Three examples of evolutionary trees, branching diagrams purporting to show the ancestor-descendant relationships among species. In these tree diagrams, species are sometimes represented by lines, as in (a), or as dots as in (b) and (c). Trees in which species are represented by dots (specified named taxa) may depict hypothesized ancestors, as in (c). Trees are discussed in detail in chapter 4.

may not represent named taxa (see chapter 4); in either case, lines are always used to signify a relationship of some kind (depending on how relationship is defined by the systematist) among the identified taxa. But lines per se may be used to symbolize a continuous ancestral-descendant lineage of one or more identified taxa, continuous in the sense that a systematist believes a sequence of change can be discerned among a series of specimens sampled from a more or less uninterrupted stratigraphic record. Lines imply continuity; it may be a conceptual continuity between dots (named, identified taxa) in which "intermediate" taxa are unknown, or the line may signify continuity (a lineage) among specimens assumed to be assignable to the same taxon. Branch points represent a speciation event; they may be named (say, with a dot), indicating an identified taxon ancestral to the two or more descendant lineages, or unnamed, indicating a hypothetical speciation event with the ancestral species unspecified. In all trees figured in this chapter, taxa are represented by dots and their relationships by lines.

As many paleontologists have stressed for years, continuously sampled stratigraphic sequences are a rarity; thus, in most cases, trees consist of lines purporting to show the hypothesized ancestral-descendant relationships among taxa perceived to be discrete. With respect to classification, the important questions are, first: how are named taxa of the tree to be subdivided and allocated to larger groups? And second: what categorical rank shall be given to these groups? We will first discuss the question of subdividing the tree.

Paleontologists since the time of Darwin have noted the arbitrary nature of subdividing a continuous lineage of evolving populations for classificatory purposes. The problem was important to Darwin and has remained so for succeeding generations of paleontologists because it focuses on the origin and definition of species. How can something continuous—for example, a gradually evolving lineage—be subdivided objectively? Of course, it cannot:

> Supposing B and C to be two species, and a third, A, to be found in an underlying bed; even if A were strictly intermediate between B and C, it would simply be ranked as a third and distinct species, unless at the same time it could be most closely connected with either one or both forms by intermediate varieties . . . unless we obtained numerous transitional gradations, we should not recognize their relationship, and should consequently be compelled to rank them all as distinct species. (Darwin 1859:297)

Darwin is suggesting here that a continuous lineage could not be subdivided, that, given continuity, only a single taxon is recognizable, and that gaps are necessary for subdivision. This belief in an arbitrary subdivision of evolutionary-phylogenetic trees persists, even though most systematists deal not with "continuous" lineages, but with discrete taxa connected by lines, lines that also branch. And because trees branch, two kinds of relationship are envisioned:

> . . . among successive taxa in an ancestral-descendant lineage, and among contemporaneous taxa of more or less distant common origin. In accordance with the usual coordinates of tree representation, the former relationships are called *vertical* and the latter *horizontal*. . . . In

dealing with recent animals or with contemporaneous faunas of fossils, only horizontal relationships are *directly* involved. . . . In temporal sequences of fossils vertical relationships are directly presented. (Simpson 1961:129; italics in original)

The dilemma for the paleontologist is resolving these two kinds of relationship within the framework of the Linnaean hierarchy:

Examination of any extensive tree or dendrogram at once reveals that classification by either vertical or horizontal relationships alone is absolutely impossible. . . . In translating the phylogeny into taxa a compromise must somewhere be effected; some divisions among taxa must be horizontal and some vertical. Choice as to just how to effect the compromise is part of the art of taxonomy, a matter of taste and ingenuity. (Simpson 1961:130)

How is the "art" of classification to be applied to trees so that nested sets of taxa, and thus a Linnaean hierarchy, can be formed? Various criteria have been suggested (Simpson 1961; Mayr 1969): (1) Temporal sequences of taxa are rarely continuous—gaps exist (that is, morphologically and temporally intermediate taxa are unknown), and these can serve as convenient, arbitrary boundaries. (2) Taxa can be clustered so that the sizes of groups of the same rank are not especially disparate. (3) Taxa can be clustered so that morphological diversity within groups of the same rank is more or less the same. And (4) taxa that are morphologically distinct or are thought to have evolved into a way of life significantly distinct from their sister-taxon can be classified apart from that sister-taxon (this may pertain primarily to the issue of ranking, but frequently involves the procedure of clustering). These are the main criteria used to subdivide trees as suggested by some well-known systematists. The first two apply to topographical aspects of trees, whereas the latter two call for additional information not capable of expression within the logical structure of a tree (e.g., Simpson 1961:113). The resolution of this last problem will be discussed in more detail shortly (see also the earlier discussion, pp. 171–75).

The question of how a tree is to be subdivided is thus answered: the procedure is unspecifiable and dependent on precedent and the "arbitrary" decision of the investigator. Given a tree, it can be subdivided in numerous ways, resulting in many classifications. Examples of some trees and their possible subdivision, at least at the

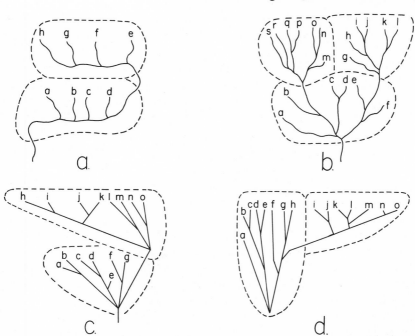

Figure 5.5 The ways in which evolutionary trees are often subdivided for classificatory purposes by evolutionary systematists (see text). (From Simpson 1961:119 and 206.)

higher taxonomic levels, are given in figure 5.5 (all from Simpson 1961). In these trees, internodes are not named; that is, ancestors are not specified. One obvious consequence of following the criteria of Simpson (1961) and Mayr (1969) for subdividing trees is the construction of not-A groups—sets of taxa clustered characteristically by the possession of shared primitive characters (symplesiomorphies). The trees of figures 5.5a and b can be assumed to exemplify this situation. In figure 5.5a taxa a–d are grouped together, the implication being that they lack those characters defining the lineage of taxa e–h. The subdivisions of figure 5.5b, illustrated by Simpson, imply a similiar situation: taxa a–f are presumably morphologically primitive to more advanced and well-defined groups g–l and m–s.

As noted in the introduction to this chapter, the history of classification has been, to a great extent, the history of eliminating not-A sets. Although many not-A sets have been rejected and are no longer in use, e.g., Radiata or Vermes, others, such as Invertebrata, Pisces, or Reptilia, can still be found in some modern classifications.

To some systematists the recognition of not-A sets is a matter of little or no concern. However, in accepting not-A sets the power, utility, and logical structure of the Linnaean hierarchy is sacrificed, if not destroyed altogether. As developed earlier, the Linnaean system of classification can represent only a nested pattern of taxa. The recognition of not-A sets assures that the nested sets of taxa within the classification cannot be translated back to the original tree. The tree and classification are independent of one another, and the classification therefore loses or misrepresents the information content of the tree.

What information content contained in trees might be useful for classification purposes? First, trees (as defined in this book; chapter 4) reflect the nested synapomorphy patterns of the taxa. Second, trees contain statements (hypotheses) about the identification of ancestors.[5] In this regard, trees are appropriate only for the depiction of hypothesized ancestral-descendant relationships among species. There do not seem to be additional kinds of information that can be expressed by the topographical structure of the tree itself (Nelson 1973c). Some workers (e.g., Mayr 1969) have suggested that the lengths of the lines and the angles of the branches can express relative degrees of divergence, but for present purposes it can be asserted that these kinds of information are not organized hierarchically within the context of a tree, and therefore are not applicable to the issue of constructing Linnaean hierarchies from trees (the "solution" of Mayr and other evolutionary systematists will be taken up shortly).

So, like cladograms, the structure of trees manifests nested synapomorphy patterns, but unlike cladograms trees propose to make

5. It is important to point out that the paleontological concept of trees just discussed is seldom as rigorously defined as in chapter 4. First, in traditional paleontological (or neontological) trees, it cannot always be assumed that construction is based on a nested pattern of synapomorphies. On the contrary, the tree itself may be a vague schematic representation of stratigraphic position and general overall similarity. Second, traditional trees often are unclear as to whether lines are meant to represent actual species continua (i.e., the line itself is named) or some symbolic conception of relationship. Here, we specifically consider taxa to be discrete units, either terminal (descendants) or nonterminal (ancestral), and lines to be representations of hypothesized phylogenetic relationships.

statements about ancestors. How are these two aspects of trees to be contained in Linnaean classifications?

Leaving aside for the moment the substantial difficulties, both scientific and philosophical, of recognizing ancestral species (chapter 4), the conversion of trees to classifications can be considered. Figure 5.6a depicts a tree in which ancestors are specified but not at internodes, hence common ancestors are not identified. The synapomorphous relationships of this tree are easily expressed in the form of a classification:

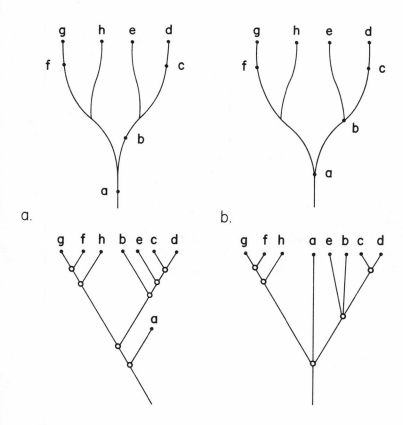

Figure 5.6 The top two diagrams are evolutionary trees in which ancestors are specified (dots). (In the left-hand tree, ancestors are not at internodes, whereas in the right-hand tree they may be.) The lower two diagrams are cladograms expressing the synapomorphic information of the trees. (See text for details.)

TAXON: a,b,c,d,e,f,g,h
  TAXON: a
  TAXON: b,c,d,e,f,g,h
    TAXON: b,c,d,e
      TAXON: b
      TAXON: c,d,e
        TAXON: e
        TAXON: c,d
    TAXON: f,g,h
      TAXON: f,g
      TAXON: h

However, if it is desired to convert this classification back into the tree from which it is derived, i.e., if we want to express the information content of the classification, the classification-tree relationship is seen to be asymmetrical. What is obtained in this process is the cladogram of the original tree (also shown in figure 5.6a, bottom). In this classification, species c is classified with species d because the two share one or more synapomorphies not shared with species e; but within the context of a traditional Linnaean hierarchy it is not possible to discern whether c is the ancestor of d, d of c, or whether neither is an ancestor. The important point, of course, is that the specified ancestor, species c, is classified with its descendant, species d, because both possess an evolutionary novelty setting apart their lineage from all others of the tree.

At this point, a convention might be agreed upon and introduced whereby ancestors could be identified within the framework of the Linnaean hierarchy. For example, let names of ancestors be placed in parentheses:

TAXON: a,b,c,d,e,f,g,h
  TAXON: (a)
  TAXON: b,c,d,e,f,g,h
    TAXON: b,c,d,e
      TAXON: (b)
      TAXON: c,d,e,
        TAXON: e
        TAXON: (c),d
    TAXON: f,g,h
      TAXON: h
      TAXON: (f),g

The original tree, figure 5.6a, can be faithfully reproduced from this classification; consequently the latter accurately represents the information content of the tree. However, this requires the introduction of a convention (in this case, the use of parentheses) not traditionally considered a component of Linnaean classification systems.

What of a tree in which common ancestors are recognized, as in figure 5.6b? The traditional classification might be as follows:

TAXON: a,b,c,d,e,f,g,h
   TAXON: a
   TAXON: b,c,d,e
      TAXON: b
      TAXON: e
      TAXON: c,d
   TAXON: f,g,h
      TAXON: h
      TAXON: f,g

Once again, ancestors are concealed within the traditional Linnaean system, and the most such a classification can do is reflect the structure of the cladogram of the tree (figure 5.6b, bottom diagram). However, as in the previous example a convention might be adopted to identify ancestors within the Linnaean hierarchy. In this case, let a parenthesis which encloses a taxon sharing equal rank with two other taxa identify the common ancestor of those taxa, but if the taxon in parenthesis shares equal rank with only one other taxon, then the former is to be considered the ancestor of the latter:

TAXON: a,b,c,d,e,f,g,h
   TAXON: (a)
   TAXON: b,c,d,e
      TAXON: (b)
      TAXON: e
      TAXON: c,d
         TAXON: (c)
         TAXON: d
   TAXON: f,g,h
      TAXON: h
      TAXON: f,g
         TAXON: (f)
         TAXON: g

There are several generalizations contained in these two examples. For any tree, those aspects of its structure that can be expressed by a Linnaean classification are precisely those aspects expressed in the corresponding cladogram, i.e., the nested pattern of synapomorphy exhibited by the taxa. Within traditional Linnaean classification schemes, concepts of ancestry and descent cannot be stored or retrieved. Such concepts cannot by themselves be treated hierarchically, but with the addition of a variety of conventions, ancestors can be notated within a Linnaean classification. One might conclude, then, that a cladogram is a more basic, more general branching diagram than a tree. The hierarchical structure of a cladogram is identical to that of a Linnaean classification system; the two are isomorphic.

All of the above assumes, of course, that ancestors can be identified or specified, for without this assumption it would seem preferable to base classification on a cladogram. We have already noted (chapter 4) the difficulty inherent in testing hypotheses of ancestry and descent.

Up to this point, the concept of trees being discussed is that most often identified with paleontological systematists (perhaps best characterized by the writings of G. G. Simpson) with their concern for "horizontal" and "vertical" relationships. The trees themselves generally attempt to depict an accurate representation of the sequence of phylogenetic branching. The classifications derived from these trees do not necessarily, as Simpson admits, reflect this branching sequence in a precise manner. Brundin (1966) and Nelson (1972b) have correctly noted that "horizontal" relationships do not exist as a separate category of kinship relationships and that what Simpson refers to are instead "similarity" relationships. Simpsonian classifications tend to emphasize "vertical" relationships—that is, genealogical or cladistic—but deviate from them when subjective assessments of "similarity" relationships are desired.

Another group of systematists—neontological evolutionary systematists—differ in that their concept of relationship emphasizes "similarity relationships" rather than kinship relationships (see Nelson 1972b, for an extended discussion):

> It is no longer legitimate to express relationship in terms of genealogy. The amount of genetic similarity now becomes the dominant consideration for the biologist.
>
> When an evolutionary taxonomist speaks of the relationship of various taxa, he is quite right in thinking in terms of genetic similarity, rather than in terms of genealogy. (Mayr 1969:79)

This concept of "genetic similarity" has never been defined precisely by the evolutionary systematists and for obvious reasons—the genetic systems of the taxa are never the subject of investigation. Rather, "genetic" similarity is estimated, and the process of estimation is based on "weighted" morphological similarity: *"weighting, then, can be defined as a method for determining the phyletic information content of a character"* (Mayr 1969:218; italics in original). Mayr (1969:220–21) lists a number of criteria used to determine the taxonomic "weight" of characters: complexity, joint possession of derived characters, constancy through large groups, characters not associated with an "ad hoc" specialization, characters not affected by "ecological shifts," and correlated suites of characters. Most importantly for this discussion, one is never told precisely how these criteria are used to construct a tree; as with the statement by Simpson noted earlier, subjective "artistic" interpretations are considered a normal aspect of taxonomic procedure (some of the problems of weighting characters in cladistic studies were discussed earlier; see pp. 66–67.

This methodology produces an ambiguous concept of a tree: "In an orthodox phyletic diagram [see figure 5.7] three kinds of information are conveyed, degree of difference in the abcissa, geological time in the ordinate, and degree of divergence by the angle of divergence" (Mayr 1969:255–56). Actually, some other kind of informa-

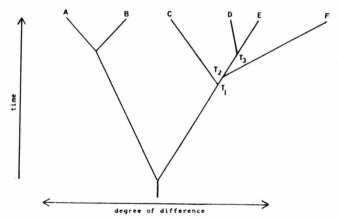

Figure 5.7 Mayr's conception of phylogeny (an evolutionary tree with terminal taxa) as expressed in a branching diagram (see text). (From Mayr 1969:256. Copyright © 1969, and used by permission of McGraw-Hill Book Company.)

tion must also be contained in the tree, for what, after all, is meant by the branching sequence? Because the methods of evolutionary systematists are *designed* to estimate "genetic" (= "weighted morphological") similarity and not cladistic relationships, *it cannot be assumed the branching sequence conveys information about phylogenetic relationships.* In this regard, "Mayrian" trees (Nelson 1972b) can be viewed as intrinsically similar to phenograms; the branching sequences attempt to reflect relative degrees of similarity.

All of this creates fundamental problems for converting "Mayrian" trees into classifications, for although it would be possible to mirror the branching sequence of such a tree in a Linnaean classification, only nested sets of taxa comprise the information content of those classifications, not a measure of similarity. Furthermore, Mayrian classifications do not mirror Mayrian trees because other criteria, such as those advocated by Simpson, are employed to subdivide the tree; consequently, not-A sets are formed.

In summary, the important aspect of trees that has traditionally attracted biologists is the specification of ancestral taxa. But if this information is to be contained within Linnaean classifications, then some ad hoc conventions must be adopted in order to identify ancestral taxa within the hierarchy. Only then is it possible to specify fully the information content of the classification and reconstruct, precisely, the original branching diagram. Only through this convention can the classification and branching diagram be isomorphic.

If this were the only problem with using evolutionary trees for classification, then the major issue would be whether the tree concept is amenable to scientific analysis. But neontological and paleontological evolutionary systematists go far beyond the question of accommodating the isomorphy between trees and classifications and propose a set of procedures for subdividing trees that in fact destroys the isomorphy itself. What these workers perceive as the strength of their approach—greater flexibility, more artistic and subjective control over the process of classification—is actually its greatest weakness, for by destroying the isomorphic relationship between trees and classifications, they destroy the information content of both; that information is, of course, the nested sets of taxa. Moreover, this approach also counteracts 200 years of systematic aspirations: the elimination of not-A sets.

The Question of Ranking: A Critique of Classical Evolutionary Classification

Linnaean hierarchies are so structured that named taxa are nested by means of categorical ranks such as order, family, genus, and species, to name only a few. Whereas all species must have a genus and species rank (thus the species taxon name is a binomial), it is not a requirement of the International Code of Zoological Nomenclature that a species be included in a suprageneric category; it is conceivable, for example, that a species may be considered of uncertain status and therefore not be classified as to family or order. What this means, of course, is that the relationships of the species are uncertain, its set-membership within a cladogram not having been resolved satisfactorily. But most species are classified within a family, order, class. etc., and it is the question of how sets of species are to be assigned a rank that has engendered much of the vehement argumentation in the systematic literature. The crux of the problem is the extent to which ranking, and thus the Linnaean hierarchy itself, is to be used to depict accurately the set-membership of the taxa in a branching diagram.

In an earlier section, it was seen how ranking can be utilized to depict set-membership, and it has been argued throughout this chapter that the logical structure of both Linnaean hierarchies and cladograms are compatible with one another only when the set-membership between the two is isomorphic. Within systematics, then, the question of ranking has been debated primarily from two different viewpoints. First, should we require that the ranking system of Linnaean classifications be used to depict precisely the set-membership of a cladogram, or should that set-membership be disregarded in the name of "equally important biological considerations"? Second, if in fact the Linnaean ranking system should depict faithfully the set-membership of cladograms, is it actually possible to classify the known diversity of organisms in this manner? This section will be organized around the debate over the first of these two questions; the second question will be examined in the section immediately following. This chapter has been an extended discussion supporting the notion that Linnaean classifications should be strictly isomorphic with the branching diagrams from which they are derived. If this is not done, then the information content of the classification is sacrificed to a greater or lesser extent. In contrast to this view, there are systematists who advocate procedures of ranking that do not lead to isomorphic set-membership of taxa between the classification and branching diagram. It seems necessary, therefore, to consider their position in some detail.

The process of classification, as Mayr (1974:113) has noted, "consists essentially of two steps, 1. the *grouping* of lower taxa (usually species) into higher taxa, and 2. the assignment of these taxa to the proper categories in the taxonomic hierarchy (*ranking*)" (italics in original).

The methods for grouping taxa were discussed in the chapter on cladogram analysis and are not of primary concern here; indeed, in recent years evolutionary systematists have shown a tendency to accept the use of synapomorphy to cluster sets of species (see chapter 1). However, philosophical differences over acceptable approaches to ranking have consistently led to conflict.

Evolutionary systematists reject the idea of an isomorphic relationship between the set-membership of branching diagrams (phylogenetic hypotheses, to them) and classifications. This viewpoint follows from their belief that classifications should be constructed so as to reflect a variety of information about the included organisms, and not just the set-membership of the taxa as expressed in the cladogram or phylogenetic diagram. For example,

> Organisms are classified and ranked, according to this theory [of the evolutionary systematists], on the basis of two sets of factors, 1. phylogenetic branching ("recency of common descent," retrospectively defined), and 2. amount and nature of evolutionary change between branching points. The latter factor, in turn, depends on the evolutionary history of the respective branch, e.g., whether or not it has entered a new adaptive zone and to what extent it has experienced a major radiation. The evolutionary taxonomist attempts to maximize simultaneously in his classification the information content of both types of variables (1 and 2 above). (Mayr 1974:95)

> If classification is based on evolutionary theory, a natural classification must be in agreement with evolutionary theory and with the whole of evolutionary theory, including all laws, mechanisms of change and subfactors thereof. By the whole of evolutionary theory, I mean all factors and mechanisms. In particular, I mean all studies of function, biological role, behavior and environmental factors required to understand the evolutionary mechanisms. And I mean all mechanisms

including that of phyletic evolution, especially all factors of the formation of genetical mechanisms and of natural selections [*sic*] underlying evolution in single phyletic lineages. Speciation is not the only
evolutionary mechanism of importance to classification contrary to
statements of many phylogenetic systematists. (Bock 1977:864)

Evolutionary classification is a system of taxa arranged in a Linnaean
hierarchy. The taxa must be monophyletic *in the sense of Simpson* and
their rank reflects the attempt to maximize simultaneously the two
semi-independent variables of amount of phyletic change as reflected
in the degree of similarity and the phylogenetic sequence of events as
reflected in the pattern of phylogenetic branching and ultimately the
pattern of speciations. *No one-to-one relationship or correlation exists
between the evolutionary classification and the phylogenetic diagram
of the groups contained in the classification.* (Bock 1977:869; italics
added)

Evolutionary classification takes into account: degrees of homologous
resemblance in *all* available aspects; the most probable phylogenetic
inferences from all data (including the foregoing resemblance plus
evolutionary analysis and weighting of the various characteristics);
and also the practical needs of discussion and communication. (Simpson 1963:25; italics in original)

The criteria used by evolutionary systematists to determine the
ranks of taxa are not absolute, nor can they necessarily be applied
objectively in any specific case. Classifications, in their view, must
be approached as "an art with canons of taste, of moderation, and of
usefulness" (Simpson 1961:227). But criteria for assigning rank have
been suggested; Mayr (1969:233), for example, lists five: (1) distinctness, or size of gap between two or more sets of species; (2) evolutionary role, or uniqueness of the adaptive zone occupied by the set
of species under consideration, particularly compared to that of
close genealogical relatives; (3) degree of difference between sets
of species; (4) the size of the taxon; and (5) the equivalence of ranking in genealogically related taxa. To this list might be added Simpson's (1963:27) call for "linguistic convenience" as a basis for establishing rank.

The scientific merit of this approach to the question of ranking
must be judged in terms of their views of the functions of biological
classification. Merit cannot be decided in terms of the isomorphic
relationship between a branching diagram (phylogenetic hypothesis) and the classification, for evolutionary systematists have
adopted, by definition, the position that this relationship in fact

frequently obscures functions of classification more important than conveying information about genealogy. An analysis of their views must focus on their own claims regarding first, the purpose of biological classification; second, the ways in which their procedures of ranking allow them to fulfill that purpose; and third, whether or not these methods accomplish the purpose for which they are intended.

Mayr (1969:229), like most evolutionary systematists, views classification as a system for the storage and retrieval of information, and in their system information is to be evolutionary in content: "The complex relationship of species with each other and the varying rates of branching and divergence must be translated into a system of taxa, conveniently ranked in the appropriate categories, in such a way that it forms as efficient as possible a system of information storage." Thus, to Mayr, a classification must meet two objectives: grouping close relatives (determined on the basis of weighted overall similarity and the five criteria listed above) and facilitating information retrieval.[6]

Evolutionary systematists assert that preferred classifications are those which are the most useful, have higher empirical content, have greater capability for predictions (or lead to broader generalizations), and which best form the basis for comparative studies. As a general view of classification, one could hardly take issue with such idealism.

The goal of evolutionary systematics is to construct classifications that go beyond the mere expression of set-membership explicit in the branching diagram. An evaluation of the evolutionary history must be included (see Bock's comments above), and this is accomplished by criteria such as those of Mayr, listed earlier. What are the characteristics of such classifications and how might the

6. Michener (1978) also views classifications as information storage-retrieval systems. Bock (1973) talks about maximizing the information content in classifications of two "semiindependent variables": degree of genetical similarity, which is said to be correlated with phenetic similarity ("greater phenotypical similarity implies greater genetical similarity," p. 377), and the branching sequence of a phylogenetic diagram. Recently, Bock (1977:865) departed somewhat from this more traditional viewpoint: "Many workers believe that the information is actually stored in the classification. This is not true, nor is the classification, itself, an information storage-retrieval system. Rather the classification serves as the foundation on which an efficient information storage and retrieval system can be constructed, be it a book, the arrangement of specimens in a museum, a computer system or whatever." The view that classifications do not store information, and that one cannot retrieve something from them, will be rejected in subsequent discussions. Indeed, throughout history, classifications—whether biological or not— have universally been thought to store and express a special kind of information, namely, set-membership.

methodological criteria of evolutionary systematists be expressed in
the actual process of constructing a classification? These questions
will be examined by some examples.

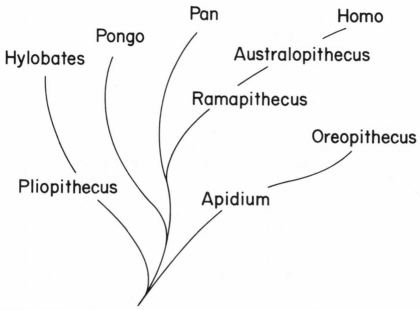

Figure 5.8 Simpson's view of hominoid phylogeny as of 1963. (After Simpson 1963:27, figure 6. Viking Fund Publications in Anthropology, no. 37. Copyright © 1963 by the Wenner-Gren Foundation for Anthropological Research, Inc., New York).

1. The hominoid classification of Simpson (1963) Figure 5.8 attempts to summarize Simpson's 1963 view of the phylogeny of the Hominoidea. The purpose here is to examine the consequences of his method of ranking, and not evaluate the validity of his taxa or their relationships. Some genera (e.g., *Kenyapithecus, Proconsul*) are omitted because it is uncertain, either from the figure or the accompanying text, precisely how Simpson would relate them to other genera. For nine genera Simpson provided the following classification:

SUPERFAMILY Hominoidea
   FAMILY Pongidae
     SUBFAMILY Hylobatinae
      Pliopithecus
      Hylobates

SUBFAMILY Dryopithecinae
Ramapithecus
SUBFAMILY Ponginae
Pongo
Pan
FAMILY Hominidae
Australopithecus
Homo
FAMILY Oreopithecidae
Apidium
Oreopithecus

The *Apidium* and *Oreopithecus* lineage was ranked as a family "because of its ancient separation plus its marked divergence from any other group now usually given family rank" (1963:26). *Hylobates* and *Pliopithecus* are classified with *Pongo* and *Pan* because all of them

> almost certainly had a common hominoid ancestry . . . and its [*Hylobates* lineage] evolutionary divergence from [*Pan* and *Pongo*] is decidedly less than that of either *Homo* or *Oreopithecus*. That would justify placing the *Hylobates* group as a subfamily of a family also containing *Pongo*, *Pan*, and some, at least, of the dryopithecine complex. Both arrangements are consistent with reasonable interpretations of the available data, and choice [between placing *Hylobates* in a separate family or with *Pan* and *Pongo*] becomes a matter of personal judgment and convenience. I continue to prefer the second alternative, partly as a matter of linguistic convenience. (pp. 26–27)

Although Simpson clearly accepts the fact that *Pan* and *Homo* are more closely related genealogically than either is to *Pongo*, he classifies *Pan* with *Pongo* and elevates *Homo* to separate familial rank because (a) "*Pan* is the terminus of a conservative lineage" and "*Homo* is both anatomically and adaptively the most radically distinctive of all hominoids," (b) "*Pan* is obviously not ancestral to *Homo*," and (c) *Homo* has diverged until it deserves family status, and if *Pan* were placed in the Hominidae, "carrying this process down will eventually require inclusion of all descendants of earlier splittings also in the latest family—eventually the whole animal kingdom would be in the Hominidae on this principle."[7]

7. It seems useful at this point to reject this last argument. Following the logic of Simpson, we could pick *any* terminal group (e.g., any Recent species) and then classify all the animal kingdom in the family of that species. More important, of course, Simpson is in error (see also 1961:165) in viewing phylogeny, and therefore classification, as proceeding from top to bot-

Simpson's approach to hominoid classification has been accepted by other evolutionary systematists, notably Mayr (1969:70): "To rank taxa according to branching points is nearly always misleading. It might necessitate, for instance, the inclusion of the African apes (*Pan*) in the family Hominidae and their exclusion from the family Pongidae."

2. The galliform classification of Mayr (1974:119) Basing his discussion on figure 5.9, Mayr argues: "There are three major families of liv-

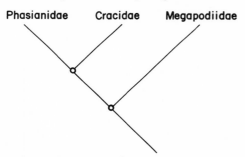

Figure 5.9 A hypothesis of relationships for three families of galliform birds.

tom, rather than vice versa (see Wiley 1978, 1979, for an extended discussion of this problem). Such views stem from the notion that it is species that evolve, not genera or taxa of higher categorical rank—a concept with which we heartily agree. But it does *not* follow that, for any monophyletic group, an ancestral species appears first, followed at some later time by genera, then families, and so forth—the view that Simpson and Mayr (quoted below) seem to support. As Wiley (1978, 1979) has pointed out, the age of a taxon depends upon the time of first appearance of the synapomorphy defining the set. The first member of the group will indeed be an ancestral species but, to be included in the set, must have at least one synapomorphy linking it to its descendants. That one (or more) synapomorphy will likewise define the higher, more inclusive set. Thus any taxon of higher categorical rank, because it is a set defined by one or more synapomorphies, "originates" at exactly the same time as its earliest included species. The confusion arises from the mistaken belief that supraspecific taxa are real entities in nature that somehow "evolve." As Wiley (1978:21) has noted: "There is no doubt that one can run from man to protist in one classificatory taxon, but, in my opinion, that taxon would be Eucaryota, not species *Homo sapiens*. There was a genus *Homo* before there was a species *Homo sapiens*, just as there was a class Vertebrata before any of the Recent vertebrates evolved. Thus, we tie together increasingly ever larger taxa on the basis of the continuum they are hypothesized to have shared in the past, and if we adopt a truly natural classification, this classification will document past continua, not bury their reality or existence."

We note also the statement of Mayr (1974:105) that phylogenetic systematists are guilty of a "fatal flaw" in reverting to Aristotle's "downward" classification rather than the "upward" classification supposedly characteristic of modern Linnaean systems. From the foregoing, it is readily apparent that this statement is incorrect: the ancestral vertebrate species, after all, existed before the ancestral primate species, and it in turn existed before *Homo sapiens*.

ing gallinaceous birds. Among these the Megapodiidae have the greatest number of primitive characters while the Phasianidae have the greatest number of derived characters. The South American Cracidae are intermediate. They share a few derived characters with the advanced Phasianidae but a far greater number of primitive characters with the Megapodiidae." On this basis, then, Mayr supports a classificatory arrangement such as the following:

> ORDER Galliformes
>   SUBORDER Cracoidea
>   FAMILY Megapodiidae
>   FAMILY Cracidae
>   SUBORDER Phasianoidea
>   FAMILY Phasianidae

In this classification Mayr would unite cracids and megapodiids on simple overall similarity, without applying, it would seem, "weighted" phenetic similarity.

3. The vertebrate classification of Romer (1962, 1966), and its advocacy by Bock (1977). Based on his studies of gill arches and other data, Nelson (1969) proposed the following genealogical classification of the Recent vertebrates using a cladogram similar to that depicted in figure 5.10 (Nelson's classification includes many other groups, including fossils, which are omitted here for simplicity):

> SUBPHYLUM Vertebrata
>   SUPERCLASS Cyclostomata
>   SUPERCLASS Gnathostomata
>   CLASS Elasmobranchiomorphi
>   CLASS Teleostomi
>   SUBCLASS Actinopterygii
>   INFRACLASS Chondrostei
>   INFRACLASS Neopterygii
>   DIVISION Holostei
>   DIVISION Teleostei
>   SUBCLASS Sarcopterygii
>   INFRACLASS Dipnoi
>   INFRACLASS Choanata
>   DIVISION Batrachomorpha
>   DIVISION Reptilomorpha
>   COHORT Sauropsida
>   SUPERORDER Chelonia
>   SUPERORDER Archosauria

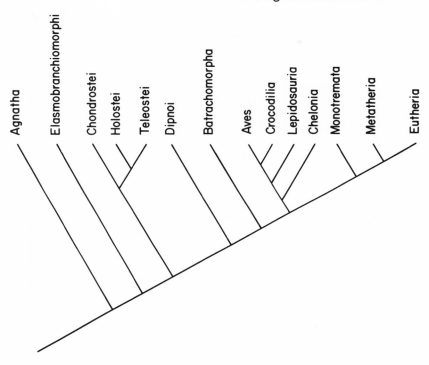

Figure 5.10 A cladistic hypothesis of the vertebrates. (After Nelson 1969.)

SERIES Crocodilia
SERIES Aves
COHORT Mammalia
SUPERORDER Prototheria
SUPERORDER Theria
SERIES Metatheria
SERIES Eutheria

Nelson's classification was critized by Bock (1977:868–69; see below for comments), who advocated adoption of the evolutionary classification of A. S. Romer (see, for example, 1962, 1966). This classification is as follows (where there are different names for the same taxon, the name in parenthesis is that of Nelson):

SUBPHYLUM Vertebrata
  SUPERCLASS Pisces
    CLASS Agnatha (= Cyclostomata)
    CLASS Chondrichthyes (= Elasmobranchiomorpha)
    CLASS Osteichthyes (= Teleostomi)
      SUBCLASS Actinopterygii
        INFRACLASS Chondrostei
        INFRACLASS Holostei
        INFRACLASS Teleostei
      SUBCLASS Sarcopterygii
        ORDER Dipnoi
  SUPERCLASS Tetrapoda (= Choanata)
    CLASS Amphibia (= Batrachomorpha)
    CLASS Reptilia
      SUBCLASS Anapsida
        ORDER Chelonia
      SUBCLASS Lepidosauria
      SUBCLASS Archosauria
        ORDER Crocodilia
    CLASS Aves
    CLASS Mammalia
      SUBCLASS Prototheria
      SUBCLASS Theria
        INFRACLASS Metatheria
        INFRACLASS Eutheria

The general pattern of branching, and even probably most of the details depicted in Nelson's cladogram, have been accepted by most vertebrate zoologists, including Romer (see especially, Romer, 1966). The classification of Romer differs from that of Nelson primarily in the elevation in rank of a number of taxa that are considered to have diverged significantly in morphology from the presumed ancestral condition, e.g., superclass Tetrapoda, class Aves, and class Mammalia.

Once again, we see this latter classification reflects evolutionary systematists' evaluation of degrees of morphological difference among taxa and the significance of the adaptive changes that are presumed to have occurred within certain lineages, to decide rank within the classificatory hierarchy. Many additional examples could have been presented, but none would have differed in principle; consequently, these three can be used to evaluate the goals and classificatory methods advocated by evolutionary systematists.

In analyzing the scientific merit of evolutionary classification, it is well to keep in mind several claims of its advocates. First, they believe classifications are designed to contain information of several diverse kinds, and that this information is manifested in the structure of classification; in other words, such information can be retrieved. Certainly they believe information should, in principle, be retrievable from classifications. Second, they believe evolutionary classifications are the best general reference system for biology because they facilitate a greater number of testable predictions and lead to more generalizations than alternative classificatory systems. The third claim is that the information being stored and retrieved includes genealogy, morphological-adaptive divergence, genetic similarity, and various other evolutionary parameters, and that such information is to be classified by judicious decisions in assigning rank.

It can be argued that the above three beliefs lead to a paradox, incapable of solution within the framework of their ideas. On the one hand, they assert that a diversity of evolutionary information must be used to construct classification, and their writings imply they believe such information is "stored" in those classifications. On the other hand, they fail to provide any rules or criteria for *retrieving* that information from their classifications, and they obviously cannot, because it would seem to be impossible. Earlier sections of this chapter demonstrated that information, to the extent it can be stored and retrieved from Linnaean hierarchies, must be expressed in terms of nested set-membership. Evolutionary systematists dream about the virtues of all possible evolutionary information (see remarks by Bock, 1977:864, cited above) being included within the structure of each evolutionary classification. It simply cannot be done. To be sure, set-membership can depict general notions of similarity relationships among species—as numerical taxonomy has demonstrated—but it has not yet been shown how the property of set-membership can convey, simultaneously, as evolutionary systematists demand, knowledge of genealogy, genetic similarity, adaptive divergence, and whatever else it is they may want to include in their classifications.

Evolutionary systematists place considerable importance on the capability of a classification for prediction and for leading to testable generalizations. Not unexpectedly, evolutionary systematists believe their classifications are superior to other approaches in this regard:

> I claim that using the criterion of a higher degree of empirical content or of testability, evolutionary classification provides the best foundation for all biological comparisons. Second, that this criterion for the best classification is the most essential distinction between classical evolutionary classification on the one hand and phenetics and cladistics on the other. (Bock 1973:381)

How are we to evaluate such a claim? Evolutionary systematists themselves seldom seem willing to engage in a dialogue about the predictive powers of classification. Rather, they are content merely to state their claim. The concept of prediction can be examined in terms of precisely what is being predicted and what renders classifications in general capable of prediction.

Evolutionary systematists, apparently universally, view prediction in terms of general overall (phenetic) similarity, that is, in the absolute numbers of similarities that might be expected:

> The common genetic program characteristic for the members of a natural taxon guarantees with a high probability that all the members of this taxon share certain characteristics. If I identify an individual as a thrush, I can make precise statements concerning its skeleton, heart, physiology, and reproduction. (Mayr 1969:79–80)

> The best classification is based on the greatest concurrence of characters in organisms which should permit the most accurate prediction of unknown characters. (Bock 1973:379)

Thus, the evolutionary systematist would argue that placing *Pan* and *Pongo* in one family, the Pongidae, and *Homo* in a separate family, the Hominidae, would allow for more predictions of unknown characters than if *Pan* were classified with *Homo,* the genus to which it is most closely related. *Pan,* they would argue, is phenotypically more similar to *Pongo* than to *Homo,* hence newly discovered features should show the *Pongo-Pan* pattern more often than not. This concept of prediction is precisely that of the numerical taxonomists. Indeed, the methodology of the latter seems superior to that of the evolutionary systematists in providing for this type of prediction. It is ironical, then, that evolutionary systematists reject the philosophy and methods that could provide the kind of prediction they seem to advocate. The solution to this problem is to modify the concept that classifications should be designed to predict overall similarity. Rather, classifications, in being consistent with their logical structure, should be capable of predicting the nested pattern of similarities observed among taxa. Viewed in this way, classifications produced by numerical taxonomy and evolutionary systematics are not maximally efficient predictors of unknown characters (Platnick 1978b).

The information content of classifications is in their nested set-membership. Classifications state that the universal set of all species under consideration can be subdivided and allocated to subsets in particular ways. Whereas one can utilize a variety of information to form those subsets, including random draw if one were so inclined, the only information to be retrieved directly from a classification is the pattern of nested sets and their membership.

How, then, can classifications predict? The pattern of nested sets provides the answer, because there is the expectation that patterns of similarities are themselves nested. This expectation seems an integral part of all taxonomic philosophies, but only the phylogenetic systematists explicitly attempt to partition the kinds of similarity into nested sets. The classifications of evolutionary systematists do not consistently partition sets of taxa according to the nested patterns of similarities, and indeed their procedure of ranking is the primary reason why such classifications are not efficient predictors. To illustrate this point, the three examples can be considered.

In order to better understand the nested pattern of taxa expressed in the hominoid classification of Simpson, the galliform classification of Mayr, and the vertebrate classification of Romer (and Bock), the set-membership of these classifications can be "retrieved" and expressed in the form of branching diagrams (figure 5.11). None of these diagrams should be interpreted as a phylogeny, but they are equivalent to the classifications in set-membership, and so are equivalent to the latter in terms of their predictive capability.

Upon comparing Simpson's phylogeny of the hominoids (figure 5.8) with the branching-diagram-equivalent of his classification (figure 5.11a), some striking differences are apparent; namely, the classification is incapable of predicting: (1) that *Apidium* and *Oreopithecus* lack certain similarities shared by all other genera; (2) that *Ramapithecus* shares certain similarities with *Australopithecus* and *Homo;* (3) that *Ramapithecus, Australopithecus,* and *Homo* share certain similarities with *Pan;* and (4) that *Ramapithecus, Australopithecus, Homo,* and *Pan* share certain similarities with *Pongo.*

This would appear to be a rather extensive list of similarities

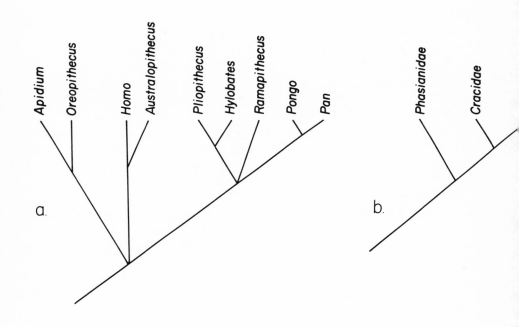

Figure 5.11 Branching diagrams depicting the set-membership expressed by the evolution

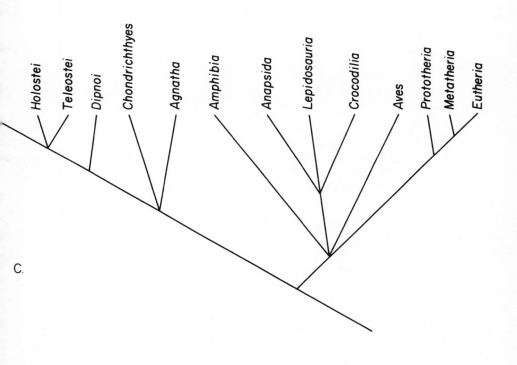

C.

ifications proposed by (a) Simpson, (b) Mayr, and (c) Romer and Bock (see text for details).

that the evolutionary classification of Simpson does not predict. Interestingly, these are quite special similarities. They are derived similarities that define four lineages within the phylogeny. If the set membership of the original branching diagram (phylogeny) were isomorphic with a classification, then the existence of these four sets of similarities would have been predicted. For example, the following classification preserves the predictions of those four sets of similarities (taxon names and absolute rank should be ignored; it is the set-membership which is important):

SUPERFAMILY Hominoidea
  FAMILY Oreopithecidae
    Apidium
    Oreopithecus
  FAMILY Hominidae
    SUBFAMILY Hylobatinae
      Pliopithecus
      Hylobates
    SUBFAMILY Homininae
      INFRAFAMILY Pongi
        Pongo
      INFRAFAMILY Homini
        TRIBE Panii
          Pan
        TRIBE Hominii
          SUBTRIBE Ramapithecininii
            Ramapithecus
          SUBTRIBE Australopithecininii
            Australopithecus
            Homo

An evolutionary systematist might well respond: "But such a classification as this could not predict that *Pan* and *Pongo* have many similarities not shared with *Homo*. The evolutionary classification does this, and is superior in this respect."

In fact, this counterargument is not entirely true. What the phylogenetic classification shown above cannot do is predict how different *Homo* may be from *Pan* and *Pongo*. But neither can the evolutionary classification. Furthermore, the phylogenetic classification indeed predicts that *Pan* and *Pongo* share a set of similarities, this prediction being made for the set of taxa included under the subfamily rank, the Homininae. In fact, probably all of the similarities that can be expected for *Pan* and *Pongo* under the evolutionary classification can also be expected under the phylogenetic classification. The difference is that the evolutionary classification makes some of its predictions at the incorrect level of the hierarchy, and this error is derived directly from the ranking procedures of evolutionary systematics. Some errors involving actual morphological similarities will be pointed out below in conjunction with the discussion of Romer's vertebrate classification.

The above arguments also apply to Mayr's classification of the galliform birds. His classification, and its derived branching diagram (figure 5.11b), does not predict the expectation of a set of similarities common to the Cracidae and Phasianidae and not shared with the Megapodiidae. However, a phylogenetic classification such as the following:

ORDER Galliformes
  SUBORDER Megapodi
    FAMILY Megapodiidae
  SUBORDER Phasiani
    FAMILY Cracidae
    FAMILY Phasianidae

would make this prediction in addition to predicting that certain features will be shared at the ordinal level, and these are precisely the features that cracids and megapodiids do share. The phylogenetic classification cannot predict how different the Phasianidae might be from the other families, but neither can the evolutionary classification.

Finally, a comparison can be made between the vertebrate classification of Nelson, on the one hand, and that of Romer (along with Bock's advocacy) on the other. In this example, the predictive capabilities of each approach become particularly clear.

Without providing any support for his assertions, Bock (1977:869) severely criticized Nelson's classification:

> Although Nelson's classification contains at one rank or another, the groups of Romer's classification, a quick comparison of the two classifications shows that the classical system advocated by Romer is far superior to that produced by Nelson in the number of possible generalizations and hypotheses that may be generated from each. Indeed some of the groups in Nelson's system are almost devoid of useful generalizations.

It can be suggested that what is needed to evaluate the merits of these two classifications is not a quick comparison, devoid of any empirical support, but a more temperate consideration of some potential empirical conflicts in the predictions of the two classifications. For example, keeping in mind the phylogenetic hypothesis on which both classifications have their foundation (figure 5.10), we find that the evolutionary classification of Romer and Bock lacks the capability to make the following predictions:

1. That there exists a set of similarities common to the Gnathostomata. Under the evolutionary classification, for example, the existence of a set of species having jaws and the associated branchial apparatus cannot be predicted, nor, apparently, is it considered a "useful generalization."

2. That there exists a set of similarities common to all vertebrates excluding agnathans and elasmobranchs.

3. That there exists a set of similarities common to the holostean and teleostean fishes.

4. That there exists a set of similarities common to the Dipnoi and the choanate vertebrates.

5. That there exists a set of similarities among the "reptiles," birds, and mammals. Thus, the evolutionary classification cannot make the prediction that there is a set of vertebrate species sharing an amniote egg.

6. That there exists a set of similarities common to birds and certain taxa of "reptiles."

This is a substantial list of predictions, involving suites of characters used by systematists for over a hundred years to define sets of species. Yet, this particular evolutionary classification does not express, in its logical structure, the capability to make these predictions.

An evolutionary systematist might make the following two-fold response to this criticism. Their first rejoinder might be predicated on the major reason for forming their classifications in the first place, namely that the phylogenetic classification of Nelson itself could not predict: (1) that there exists a set of similarities common to all fishes (Romer's Pisces); (2) that there exists a set of similarities common to the dipnoans and actinopterygian fishes; and (3) that Chondrichthyes, Osteichthyes, Aves, and Mammalia all exhibit significant morphological divergence from their presumed ancestors and have entered strikingly new adaptive zones.

The first two points can be answered very simply. All similarities predicted for Romer's Pisces are included in the similarities predicted by Nelson's Vertebrata. Romer's Pisces is an artificial grab bag set of taxa formed because they share some primitive features (and probably few at that; most likely it is their "fishiness" that prompted Romer to separate them from the land vertebrates); those same features, if they can be recognized, would be derived at the level of the Vertebrata. The second point can be answered similarly: Romer's Osteichythyes is another grab bag, and those features shared by the dipnoans and actinopterygians are primitive and predicted by the taxon Teleostomi in Nelson's classification. The third point can be rejected entirely; whereas the phylogenetic classification cannot express information about divergence or adaptation, there is nothing inherent in the logical structure of Romer's classification that makes that information any more apparent. We are compelled to interpret Romer's classification only in terms of set-membership.

The second response of the evolutionary systematist probably would take the following form:

> Because the approach of evolutionary classification does not insist on a one-to-one correspondence of the classification and the phylogeny, separate sets of hypotheses about groups are needed to cover both aspects of relationships. And it is necessary at the completion of an evolutionary classificatory analysis to present the conclusions in a formal classification and in a phylogeny. Many workers omit the phylogeny; this omission causes problems for others who may need the exact phylogeny for their studies and yet cannot obtain this information from the classification. (Bock 1977:872)

But this response places the evolutionary systematist in a precarious position logically. Their classifications are supposed to contain phylogeny as one of their "semiindependent" variables, yet they admit that it cannot be retrieved from their classifications, and therefore a phylogenetic diagram must be published along with the classification. Furthermore—and Bock seems to recognize the threads of the problem in the above citation— *neither can the second "semiindependent" variable—morphological and adaptive divergence—be extracted from evolutionary classifications without an accompanying phylogenetic diagram.* One cannot interpret their system of ranking, as a means to convey information about morphological similarity or dissimilarity, without reference to a phylogenetic hypothesis. This point is also emphasized by Farris (1977:847):

> The difficulty with attempting to use categorical rank to convey information about degree of distinctiveness of taxa is that in order for it to work the categorical level of separation between the two taxa obviously must generally be interpretable in terms of degree of difference. But *Homo* [in Simpson's classification, discussed earlier] is separated from *Hylobates* at the same categorical level as that at which *Homo* is separated from *Pan,* conveying the "information" that *Homo* is equally distinct from *Pan* and *Hylobates. Homo,* however, is admitted to be much more similar to *Pan* than it is to *Hylobates. Basing categorical rank on degree of difference is thus seen to distort or conceal the very kind of information it was intended to express* (Italics added)

Thus, it seems that neither of the two kinds of information purported to be contained in evolutionary classifications can in fact be retrieved. If a method of classification, *in itself,* without reference to all potentially accompanying diagrams, text discussion, and so forth, cannot stand alone as a general reference system, what is there to recommend it? The poverty of evolutionary systematics is not that it is a "sloppy system" because of "sloppy material" (Bock 1977:867); rather the method and theory lack logic and conceptual clarity, qualities so necessary for placing the system in order.

Conclusion

It was shown at the beginning of this chapter that a historical tendency has existed in classification since the time of Linnaeus. That tendency has been toward "natural classification" and has been manifested by a general desire of systematists to eliminate not-A sets. Thus, systematists through the years have strived to recognize strictly monophyletic taxa and to eliminate those which are nonmonophyletic.[8]

8. The literature on the possible kinds of nonmonophyly and their definitions has been tortuous to say the least (the reader is referred to Ashlock 1971, 1972; Farris 1974; Hennig 1965, 1966, 1975; Nelson 1971b, 1973b; and Platnick 1977a). Whereas most workers now accept the general concept of monophyly as formulated by Hennig "A monophyletic group is a group of species descended from a single ('stem') species, and which includes all species descended from this stem species" (Hennig 1966:73), there are still diverse opinions on how to view nonmonophyly. The term *holophyletic* has been used by Ashlock and other evolutionary systematists to refer to the concept of monophyly *sensu* Hennig and other phylogenetic systematists; because holophyletic is a synonym of monophyletic, only the latter term will be used here. There have been two aspects of nonmonophyly, paraphyly and polyphyly. Aside from the intracies of how these terms are to be formally defined and applied (we suggest those of Farris

Evolutionary systematists, it should be clear, subscribe to an assortment of methodological rules having the overall effect of reversing, or at least resisting, this historical tendency. Their procedures of ranking are specifically designed to produce nonmonophyletic taxa, i.e., not-A sets. The classifications discussed above amply demonstrate this (Pisces, Reptilia, Pongidae, and so on). It seems that little can be done, at this point, to convince established evolutionary systematists of the need to reevaluate their premises more carefully: "I do not care whether taxa are paraphyletic (reptiles) or holophyletic (birds). I readily accept both" (Michener 1978:114). On the other hand, the tendency to eliminate not-A sets is vigorous and gains increasing support from the systematic community in particular, and biology in general.

Cladograms, Trees, and Classification

Cladograms and Linnaean classification schemes share a similar logical structure: they consist of sets within sets. In cladograms, the nested sets of taxa reflect nested sets of synapomorphies; in Linnaean classifications, the nested sets of taxa are the primary information, for the classification itself could have been based on a branching diagram derived from one of a variety of approaches. Why, then, are cladograms, as branching diagrams, considered to be of value in classifying organisms?

So far, this book has developed the viewpoint that cladograms can be used to make general statements about the history of organisms. Given the acceptance of the evolutionary paradigm, clado-

---

1974; Platnick 1977a), the conception of many workers seems to be that *paraphyly* refers to taxa classified together primarily on the basis of shared primitive characters, with some other taxon genealogically related to taxa within that group ranked as a separate, higher taxon (the hominoid classification of Simpson, in which *Australopithecus* and *Homo* are not classified with *Pan,* is illustrative). *Polyphyletic,* on the other hand, has been viewed as referring to taxa classified together on the basis of sharing some presumably derived character later shown to be independently acquired (convergent). Perhaps the only consequence of this semantic distinction is that evolutionary systematists admit both monophyletic and paraphyletic groups into their classificatory system, whereas phylogenetic systematists accept only monophyletic groups. Consequently, a distinction more refined than that between monophyly and nonmonophyly hardly seems important, and only these two terms will be used in this book.

grams are hypotheses about the *structure* of that history, that is, not specifically about the history itself, but about the structure of the relationships of the organisms as expressed in their patterns of shared evolutionary novelties. This underlying structure is a more fundamental, more general aspect of evolutionary history than the numerous hypotheses that might be constructed about specific evolutionary events and, consequently, is the reason why a number of evolutionary trees may have identical *structures,* although not necessarily identical topologies. Classifications based on cladograms are immune to the various problems and special requirements engendered by alternative evolutionary trees, yet contain what is the common denominator to all alternative trees, the nested sets of taxa. It is appropriate at this point to examine the structural characteristics of trees and cladograms in some more detail and to demonstrate the relationship of that structure to classificatory hierarchies.

The Structure of Trees and Cladograms

With regard to configurations of trees, two separate topologies can be identified: (1) trees in which none of the included taxa is specified as being ancestral to any other taxon or group of taxa within the tree (i.e., all taxa are terminal), or (2) trees in which one or more of the included taxa are specified as being ancestral to some other taxon or group of taxa within the tree.

Furthermore, trees in which ancestral taxa are specified may be categorized as follows: (2a) trees in which the specified ancestor or ancestors are directly ancestral to only one other taxon; (2b) trees in which a specified ancestor or ancestors are ancestral to two or more taxa (common ancestors are specified); (2c) some combination of 2a and 2b.[9]

Given a tree of type 1 in which no ancestral taxon is specified, the tree is identical in configuration to the corresponding cladogram (figure 5.12 top). In the tree, branch points signify hypothesized speciation events; in the cladogram, branch points signify the joint possession of synapomorphy. Neither the cladogram nor the tree has to

9. In this discussion, unlike the formalization of Nelson (1973c), hypothetical ancestors will not be considered; the question of interest is the structure of known taxa, not of unknown, hypothetical entities. Despite this difference, Nelson's treatment and the one given here exhibit many similarities.

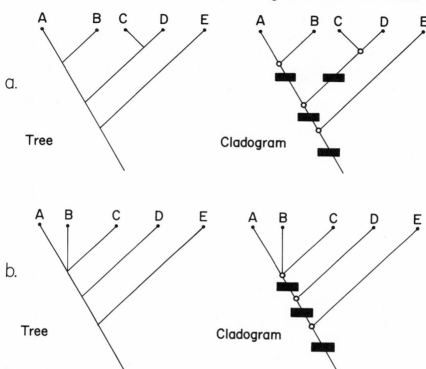

Figure 5.12 Diagrams depicting the similarities in the structures (set-membership) of trees and cladograms (see text).

be dichotomized fully in order for this isometric correspondence in structure to hold (figure 5.12 bottom). Thus, in cases in which ancestors are not specified, the structures of the cladogram and its corresponding tree are identical and can be expressed hierarchically in a classification. The classifications corresponding to the diagrams in the top and bottom of figure 5.12 are as follows:

(top)	(bottom)
TAXON ABCDE	TAXON ABCDE
TAXON ABCD	TAXON ABCD
TAXON AB	TAXON ABC
TAXON A	TAXON A
TAXON B	TAXON B
TAXON CD	TAXON C
TAXON C	TAXON D
TAXON D	TAXON E
TAXON E	

Any alternative classification for these two examples that changed the set-membership of the taxa would destroy the topological information implied by the cladogram and the tree.

In the case of Type 2a trees, none is isomorphic with the cladogram (figure 5.13); nevertheless, the underlying structure (i.e., the set-membership) of all the trees is identical to that of the cladogram. Thus, a single classificatory statement can summarize the basic structural information contained in a diversity of evolutionary trees:

```
TAXON  ABCD
 TAXON  ABC
  TAXON  AB
   TAXON  A
   TAXON  B
  TAXON  C
 TAXON  D
```

Once again, that information is the nested pattern of synapomorphy expressed by the cladogram.

Type 2b trees, in which common ancestors are specified, correspond in structure to cladograms containing one or more trichotomies (figure 5.14). In the example, taxa A, B, and C share one or more synapomorphies, but no two of the taxa share synapomorphies excluding the third. The set-membership implied by the trees is identical to that of the cladogram and, again, can be expressed in hierarchical form:

```
TAXON  ABCDE
 TAXON  ABCD
  TAXON  ABC
   TAXON  A
   TAXON  B
   TAXON  C
  TAXON  D
 TAXON  E
```

It can be concluded on the basis of these examples that cladograms are more general hypotheses about the history of life than are trees, and that, as such, there is much to recommend them as a basis for classification. Unless one wants to introduce some convention to identify ancestors within the Linnaean hierarchy, as was discussed earlier in this chapter, the logical structure of a Linnaean classification is incapable of expressing the specific kind of informa-

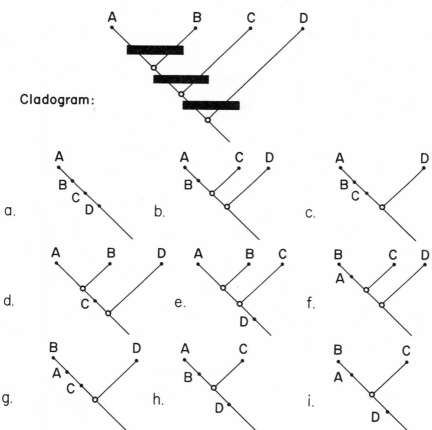

Figure 5.13 (a–i) Some possible evolutionary trees for the four species, A–D. The structure (set-membership) of all trees can be expressed by the same cladogram. (See text.)

tion unique to trees, i.e., ancestry. The information of cladograms, on the other hand, consists entirely of set-membership and this is easily translated into a scheme of Linnaean hierarchies. The remainder of this chapter is devoted to discussing the procedures necessary to produce phylogenetic classifications based on cladograms.

The Nature of Phylogenetic Classifications

Throughout this book, a distinction has been maintained between cladograms and trees. This distinction has arisen in recent years as

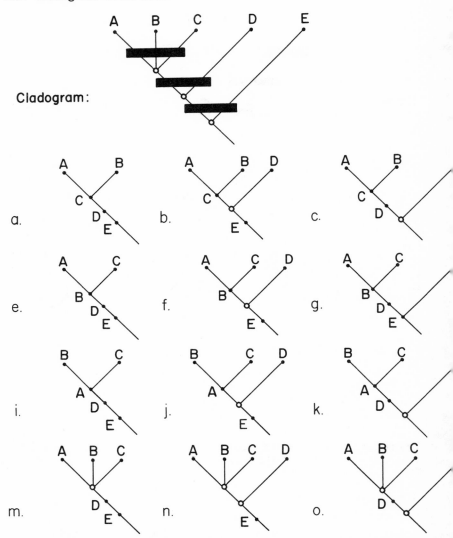

Figure 5.14 (a–p) Evolutionary trees in which common ancestors are specified. The structure (set-membership) of all trees can be expressed by the same cladogram. (See text.)

a means of separating the problems associated with analyzing pattern (e.g., the distributions of similarities and differences among taxa) from those involving a more direct consideration of evolutionary events and processes. Such a distinction seems desirable for

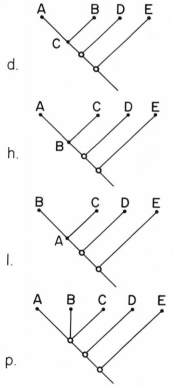

d.

h.

l.

p.

it tends to require greater analytical precision on the part of a systematist investigating phylogenetic and evolutionary questions. This is particularly true if the primary goal of the research is construction of hypotheses about evolutionary trees. How, after all, can one investigate evolutionary trees without first dealing with hypotheses about cladograms, i.e., about synapomorphy pattern?

The reader will realize, however, that phylogenetic systematists have not always recognized a conceptual difference between clado-grams and trees, and indeed some may still prefer not to do so. As a consequence, since the mid-1960s, many of these workers have viewed cladograms as trees in which ancestors are not being speci-fied: cladograms defined in this way are taken to be hypotheses about phylogenetic branching patterns and discussions about taxa are in terms of phylogenetic (genealogical) relationships. This stance poses no special problems for our discussion of classifica-tion, because both the more traditional and the newer concept of cladograms share a similar basis: both depict synapomorphy pattern, and it is that pattern, expressed by nested sets of taxa, which can serve as a basis for classification.

Given a cladogram or hypotheses of phylogenetic relationships, how are these to be expressed in a classification? The answer that follows will summarize the conventional views outlined by phylogen-etic systematists from Hennig (1966) to the present (see Brundin 1966; Nelson 1972a, 1973; Cracraft 1974b; Bonde 1977).

Both cladograms and Linnaean classifications depict hierarchi-cally arranged sets of taxa. The problem, then, is to effect a symmet-rical conversion of the cladogram to a classification in order to max-imize the information content of the latter. It has been pointed out that taxa are classified according to the two aspects of Linnaean hierarchies—subordination and sequencing—and it is these aspects which provide the basis for constructing any kind of classification, including one that is phylogenetic (genealogical). The aspect of subordination has preoccupied taxonomic thinking since the time of Linnaeus, and only recently has sequencing been proposed as a method to convey phylogenetic relationships (see below).

Given the cladogram, or phylogenetic hypothesis, shown in fig-ure 5.15, what might be a corresponding classification? Ignoring for the moment the absolute rank of the included taxa or their proper names, one possible classification might be as follows:

Subphylum Vertebrata
   Superclass Batrachomorpha
     Class Amphibia
   Superclass Amniota
     Class Archosauria
       Subclass Crocodilia
       Subclass Aves

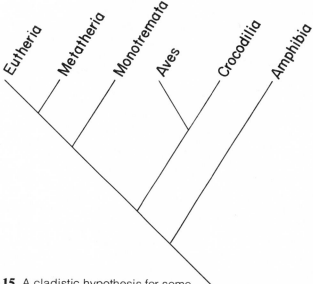

Figure 5.15 A cladistic hypothesis for some major groups of vertebrates.

CLASS Mammalia
  SUBCLASS Monotremata
  SUBCLASS Theria
    INFRACLASS Metatheria
    INFRACLASS Eutheria

This classification accurately depicts the nested sets of taxa shown in the cladogram and, consequently, also precisely expresses the implied phylogenetic relationships. Phylogenetic classifications, including this one, contain only monophyletic taxa (to the extent that we have corroborated hypotheses of relationships); that is, in a phylogenetic sense, all taxa, regardless of rank, contain all known species hypothesized to be descended from a common ancestral species. Moreover, in those classifications in which subordination is used to indicate cladistic relationships, it is implied that coordinate sister-groups stem from a common ancestral species; hence these groups are of the same absolute age. Within any phylogenetic classification *of this kind,* therefore, the categorical rank of a taxon is an indication of relative age. But to repeat, the most important consideration in constructing phylogenetic classifications is the correct depiction of the set-membership exhibited by the branching diagram (cladogram). Thus, in this example, the information content will be completely specified in the classification as long as the set-membership takes the following form:

GROUP: Amphibia, Crocodilia, Aves, Monotremata, Metatheria, Eutheria
  SUBGROUP 1: Amphibia
  SUBGROUP 1a: Crocodilia, Aves, Monotremata, Metatheria, Eutheria
    SUBGROUP 2: Crocodilia, Aves
    SUBGROUP 2a: Monotremata, Metatheria, Eutheria
      SUBGROUP 3: Monotremata
      SUBGROUP 3a: Metatheria, Eutheria

The names of the taxa or their absolute ranks are of somewhat less consequence.

In laying the foundations of phylogenetic classification, Hennig (1950, 1966) was attempting to provide principles whereby the hierarchical taxic relationships of phylogenetic hypotheses could be translated unambiguously into a classification, and vice versa. His phylogenetic hypotheses represented relationships dichotomously (to the extent that analysis so permitted); consequently a dichotomous hierarchical treatment of subordinate ranks, as just summarized and illustrated, permitted the precise expression of those relationships. Hennig's approach to classification, then, logically resulted in sister-group taxa being coordinate and of the same absolute rank—he even viewed this as "the fundamental construction principle of the phylogenetic system" (1966:156).

Much of Hennig's discussion centered around the goal of having a single classificatory system for all groups of animals. Under such a system, there would be a direct correlation between age of origin of a group and its absolute rank. Hennig also suggested that if such a system could be constructed, taxa of the same rank might be more comparable than is now possible with current classifications. Although his suggestion has been ridiculed by various systematists, Hennig himself actually took a somewhat more realistic view of current systematic practice (1966:191):

> The requirement that rank designations must express the comparability of [taxa assigned to the same] categories—however remotely related the groups—is not a fundamental principle of phylogenetic systematics to the same degree as the requirement that the system must contain only monophyletic groups and that sister groups must be coordinate and be given the same rank. This fact makes a

compromise possible . . . each specialist can erect a consistent phylogenetic system for his group without any necessity for correspondence on the basis of equivalent age between the absolute rank order of his categories and the absolute rank order of other groups of animals.

Hennig (1966) did not discuss in detail the limitations of using subordination of ranks, by itself, to convey the hierarchical information of a branching diagram. But it did not take long for systematists, including proponents of phylogenetic classification, to call attention to the problems of relying entirely on subordination. Those systematists sharing Hennig's goal of conveying phylogenetic information in classification have offered some imaginative solutions to these problems; those not sharing his goal, on the other hand, have used these problems as a reason for rejecting outright his system of classification and for simultaneously ignoring alternative solutions proposed by phylogenetic systematists.

Perhaps three major problems of Hennig's subordination scheme have been identified.

Problem 1 Because each branch of a branching diagram (phylogeny) necessitates subordination, it becomes apparent that such a scheme cannot easily accommodate the known taxonomic diversity of most groups of organisms (assuming that their phylogenetic relationships are generally accepted). Such a classification would require (a) the creation of new categorical ranks and (b) the creation of new names for taxa at each of those ranks.

Some typical comments in the literature follow:

If phylogenetic classification proceeds usually by dichotomous divisions, and very unequal ones at that, it will necessitate the use of many more categories than were needed for older, "formal" systems. (Crowson 1970:260)

Since in any dichotomous dendrogram there is one less branching point than there are terminal points, the number of names needed for a complete description of the cladogram is one less than the number of species contained in the group. . . . I don't believe that the cladists want to burden systematics with the number of names needed to make formal cladistic classifications completely describe cladograms. (Ashlock 1974:96)

It appears very doubtful that the increased expression permitted by a phylogenetic classification will compensate for the confusion and problems that will accompany the proliferation of categorical levels and taxa names. (Bock 1977:862)

The "problem" of subordinated phylogenetic classifications is illustrated in McKenna's (1975) preliminary attempt to classify phylogenetically the higher taxa of mammals. It was necessary, in this instance, for McKenna to create new categorical rank names (e.g., Parvorder, Superlegion, Legion, Sublegion, Supercohort, Magnorder, Grandorder, Mirorder), along with many new taxon names. Nevertheless, in so doing, he presented a classification reflecting phylogenetic relationships, to the extent that he discussed them.

Problem 2 The notion that coordinate sister-taxa are to have the same rank in a subordinated phylogenetic system is said to lead to objectionable consequences when classifying fossils:

> The greatest possible difficulties for phylogenetic classification which could arise from these cases would be if Recent species of *Nautilus* were traced to separate Mesozoic ancestors. . . . We might then find ourselves obliged, by the logic of phylogenetic systematics, to place in separate tribes or even subfamilies species which systematists have regarded as only barely distinguishable from each other, or to make separate families for what have been considered as poorly separated genera. These problems, however, have not yet arisen, and may never arise. (Crowson 1970:248)

> The kind of systematic presentation Hennig then had in mind would be very much open to a charge of conceptual redundancy because of a proliferation of monobasic group names. All species which originated and became extinct in the Triassic would, if orders are defined as groups originating in the Triassic, be considered to represent monobasic orders, families, and genera. (Griffiths 1973:339)

> The problem raised particularly by the inclusion of older fossils is that they are likely to engender a procession of higher and higher ranks to express the relationships of increasingly plesiomorph sister-taxa. (Patterson and Rosen 1977:155)

Hennig desired a classification encompassing all known organisms, and proposed that ranking within such a system be determined not only by relative time of origin, which would be a logical result of any hierarchical, subordinated, phylogenetic system, but also by absolute time of origin. The latter would only be possible, it would seem, if in fact biology had a complete branching system for all organisms, otherwise ranking by absolute age would be mean-

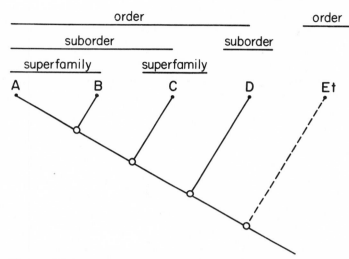

Figure 5.16 A cladogram for five taxa, four of them Recent and one fossil (see text).

ingless from group to group. As this situation is moot within biology, the question of ranking by absolute age can be safely abandoned until some future date. But the question of relative ranking still causes some concern with regard to fossil taxa. Consider, for example (figure 5.16), four Recent taxa, A–D, each, let us say, with many subgroups and species, and traditionally classified as follows:

ORDER A–D
  SUBORDER D
  SUBORDER A–C
    SUPERFAMILY C
    SUPERFAMILY A–B

A newly discovered fossil, species E, would have to be placed in its own order since it would be the coordinate sister-taxon of order A–D. It is conceivable, therefore, that the discovery of additional fossils could result in a number of higher categories being created for single fossil species or for only a few species. As Hennig notes (1966:192), however, this may or may not be considered objectionable "only insofar as it contradicts ideas associated with our more or less typological way of thinking of the higher categories." Despite this, there has been a general desire on the part of systematists to avoid classifications of this sort.

Problem 3 If ranking within phylogenetic classifications is determined by the branching sequence of a cladogram, then newly discovered taxa (Recent or extinct), or changes in our understanding of relationships, or uncertainty about those relationships in the first place, all might necessitate major alterations in the ranks of named taxa, thus creating the potential for an inherently unstable classification.

It can be argued, of course, that change in our understanding of relationships is a worthwhile reason for a change in classification, and surely few such changes tend to wreak havoc with a system of ranking. Nonetheless, it is desirable to incorporate newly discovered taxa and reallocations of systematic status into classifications with a minimum of change in rank and categorical names. One of the functions of ranking, and classification in general, is to convey information about set-membership, and in order for this to be carried out effectively, categorical and rank names should be as stable as possible.

It should be noted, then, that these problems do not directly challenge the logical structure or the capability of phylogenetic classifications to convey precise statements about set-membership. Rather, these are problems more of convenience and communication resulting from the difficulty of remembering categorical and taxic names. Moreover, they are not necessarily problems restricted to phylogenetic classifications but are shared by other classificatory systems to one degree or the other. And they are problems to the extent that systematists view current systems and their conventions as monolithic and resistant to change.

Phylogenetic systematists have proposed a number of classificatory conventions to circumvent the problems just noted. Hennig (1969) himself suggested replacing categorical names with a numbering system (table 5.1); the resultant classification is hierarchical but not Linnaean. His classification maintains a dichotomous arrangement of coordinate sister-groups, but the numbering system has several disadvantages. First, the numbers are functioning like categorical rank names and, as long as the classification is dichotomous, any changes that might create problems for these Linnaean ranks would appear to do the same for number ranks. Second, placed within the context of our systematic tradition, numbers are not as effective for verbal or written communication as the pre-existing

Table 5.1 Hennig's Systematization of the Insecta[a]

1. Entognatha	2.2.2.2..3.2..2. Condylognatha
1.1. Diplura	2.2.2.2..3.2..2.1. Thysanoptera
1.2. Ellipura	2.2.2.2..3.2..2.2. Hemiptera
1.2.1. Protura	2.2.2.2..3.2..2.2.1. Heteropteroidea
1.2.2. Collembola	2.2.2.2..3.2..2.2.1.1. Coleorrhyncha
2. Ectognatha	2.2.2.2..3.2..2.2.1.2. Heteroptera
2.1. Archaeognatha (Microcoryphia)	2.2.2.2..3.2..2.2.2. Sternorrhyncha
2.2. Dicondylia	2.2.2.2..3.2..2.2.2.1. Aphidomorpha
2.2.1. Zygentoma	2.2.2.2..3.2..2.2.2.1.1. Aphidina
2.2.2. Pterygota	2.2.2.2..3.2..2.2.2.1.2. Coccina
2.2.2.1. Palaeoptera	2.2.2.2..3.2..2.2.2.2. Psyllomorpha
2.2.2.1..1. Ephemeroptera	2.2.2.2..3.2..2.2.2.2.1. Aleyrodina
2.2.2.1..2. Odonata	2.2.2.2..3.2..2.2.2.2.2. Psyllina
2.2.2.2. Neoptera	2.2.2.2..3.2..2.2.3. Auchenorrhyncha
2.2.2.2..1. Plecoptera	2.2.2.2..3.2..2.2.3.1. Fulgoriformes
2.2.2.2..2. Paurometabola	2.2.2.2..3.2..2.2.3.2. Cicadiformes
2.2.2.2..2.1. Embioptera	2.2.2.2..4. Holometabola
2.2.2.2..2.2. Orthopteromorpha	2.2.2.2..4.1. Neuropteroidea
2.2.2.2..2.2..1. Blattopteriformia	2.2.2.2..4.1..1. Megaloptera
2.2.2.2..2.2..1.1. Notoptera (Grylloblattodea)	2.2.2.2..4.1..2. Raphidioptera
2.2.2.2..2.2..1.2. Dermaptera	2.2.2.2..4.1..3. Planipennia
2.2.2.2..2.2..1.3. Blattopteroidea	2.2.2.2..4.2. Coleoptera
2.2.2.2..2.2..1.3.1. Mantodea	2.2.2.2..4.3. Strepsiptera
2.2.2.2..2.2..1.3.2. Blattodea	2.2.2.2..4.4. Hymenoptera
2.2.2.2..2.2..2. Orthopteroidea	2.2.2.2..4.5. Siphonaptera
2.2.2.2..2.2..2.1. Ensifera	2.2.2.2..4.6. Mecopteroidea
2.2.2.2..2.2..2.2. Caelifera	2.2.2.2..4.6..1. Amphiesmenoptera
2.2.2.2..2.2..2.3. Phasmatodea	2.2.2.2..4.6..1.1. Trichoptera
2.2.2.2..3. Paraneoptera	2.2.2.2..4.6..1.2. Lepidoptera
2.2.2.2..3.1. Zoraptera	2.2.2.2..4.6..2. Antliophora
2.2.2.2..3.2. Acercaria	2.2.2.2..4.6..2.1. Mecoptera
2.2.2.2..3.2..1. Psocodea	2.2.2.2..4.6..2.2. Diptera

[a]From Hennig, 1969.

categorical rank names (although the taxon names are the same in both systems). Finally, given a sufficiently large classification, the number ranks preceding the taxon name may be inordinately long, thus further reducing its effectiveness in communication. Hennig's replacement of rank names with numbers was undertaken in the apparent belief that dichotomous subordination was the only means of conveying phylogenetic relationships. It therefore seems appropriate at this point to consider some of the alternative methods for forming phylogenetic classifications within the Linnaean hierarchical system, methods that do not rely necessarily only on subordination.

Phyletic Sequencing

The topology of Linnaean hierarchies has two aspects, subordination of ranks and the linear listing of taxon names of the same rank. Until recently this second aspect was not used to convey information about set-membership but Nelson (1972b) showed that it can be so utilized. Phyletic sequencing, as this can be called in order to distinguish it from conventional listing in which information about relationships is not implied, has been discussed and applied by a number of workers (e.g., Cracraft 1974b; Wiley 1976; Patterson and Rosen 1977; Bonde 1977). Phylogenetic relationships (or set-membership in its most general form) are conveyed by sequencing using a simple convention: the first taxon in the sequence is considered the sister-taxon of all taxa listed below it, the second taxon the sister-taxon of all below it, and so on. As a simple example, consider the branching diagram or phylogeny of figure 5.17. Using the phyletic

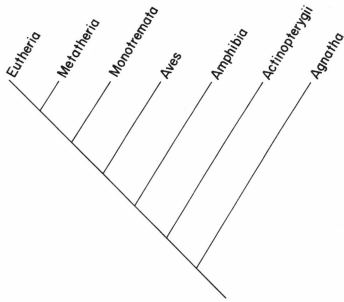

Figure 5.17 A cladistic hypothesis for some major groups of vertebrates.

sequencing convention this branching diagram can be specified precisely by the following classification:

GROUP 1: Agnatha
GROUP 2: Actinopterygii
GROUP 3: Amphibia
GROUP 4: Aves
GROUP 5: Monotremata
GROUP 6: Metatheria (or Eutheria)
GROUP 7: Eutheria (or Metatheria)

This classification signifies that the Agnatha are the sister-group of the six taxa listed below it, the Actinopterygii the sister-group of the five taxa below it, and so on. Rank was not specified; its choice would have no effect on the information content of the classification. Given only these taxa, one rank is sufficient to specify relationships; compare this to a conventional, subordinated phylogenetic classification in which, minimally, six ranks are required:

SUPERCLASS Agnatha
SUPERCLASS Gnathostomata
  CLASS Actinopterygii
  CLASS Choanata
    SUBCLASS Amphibia
    SUBCLASS Reptilomorpha
      INFRACLASS Aves
      INFRACLASS Mammalia
        DIVISION Monotremata
        DIVISION Theria
          COHORT Metatheria
          COHORT Eutheria

In a classification in which taxa are sequenced only, absolute rank would be determined by (a) the rank given to the group as a whole and (b) an arbitrary choice of rank for the included taxa as long as that rank is lower than that applied to the group as a whole. Thus, if the seven taxa as a group were classified within the subphylum Vertebrata, then, in a sequenced classification, each could be given a rank as high as superclass or as low as, say, family; and rank would have no influence on information content.

Phyletic sequencing by itself solves one of the major problems identified for purely subordinated classifications, namely, the proliferation of categorical and taxon names.

If subordination and phyletic sequencing are combined, there exists the potential for many different classifications, all of which specify the same set-membership. For example, two of these classifications might be as follows:

SUPERCLASS Agnatha
SUPERCLASS Gnathostomata
  CLASS Actinopterygii
  CLASS Amphibia
  CLASS Aves
  CLASS Mammalia
    SUBCLASS Monotremata
    SUBCLASS Metatheria (Eutheria)
    SUBCLASS Eutheria (Metatheria)

SUPERCLASS Agnatha
SUPERCLASS Gnathostomata
  CLASS Actinopterygii
  CLASS Choanata
    SUBCLASS Amphibia
    SUBCLASS Aves
    SUBCLASS Monotremata
    SUBCLASS Metatheria (Eutheria)
    SUBCLASS Eutheria (Metatheria)

When phyletic sequencing and subordination are used in this way, considerable subjectivity enters into the choice of ranks, the number of subordinations, and the number of taxa to be sequenced. Any nonterminal dichotomy will necessitate subordination; for instance, if Crocodilia had been included as the sister-group of birds, then Aves and Crocodilia would need to be subordinated below some other taxon name (e.g., Archosauria). Thus, using subordination and phyletic sequencing, rank will be determined by (a) the initial rank given the group as a whole, and (b) an arbitrary choice of ranks, lower than the initial rank, and depending on the scheme of subordination and sequencing.

It has been suggested (Cracraft 1974b) that a combination of sequencing and subordination might serve as a bridge between the goals of the phylogenetic systematists on the one hand, and evolutionary systematists on the other. Such classifications are strictly phylogenetic, i.e., they admit only monophyletic taxa and the phylogeny can be specified precisely; this is certainly the goal of phylogenetic classification. Moreover, the freedom given to the systematist regarding choice of rank and the pattern of sequencing and subordination would permit an evolutionary systematist to express a general statement about the degree of divergence of certain taxa and still preserve the phylogenetic relationships. To illustrate this point, Simpson's (1963) phylogeny of the hominoid primates (figure 5.8) might be classified as follows using subordination and phyletic sequencing:

    SUPERFAMILY Hominoidea
     FAMILY Oreopithecidae

> *Apidium*
> *Oreopithecus*
> FAMILY Hylobatidae
> > *Pliopithecus*
> > *Hylobates*
> FAMILY Pongidae
> > *Pongo*
> FAMILY Hominidae
> > SUBFAMILY Paninae
> > > *Pan*
> > SUBFAMILY Ramapithecinae
> > > *Ramapithecus*
> > SUBFAMILY Homininae
> > > *Australopithecus*
> > > *Homo*

Alternatively, the Paninae, Ramapithecinae, and/or Homininae could be raised to family rank.

An evolutionary systematist thus may believe that such a classification has an advantage over the phylogenetic classification presented earlier (p. 206) in that *Australopithecus* and *Homo* are now classified at a higher rank (subfamily, or even family) than before (subtribe); classifying *Pan* with *Homo* instead of *Pongo* emphasizes their genealogical relationships and, of course, their closer genetic similarity. On the other hand, even though some subjective element might be used to express degree of difference in such a sequenced classification, it should be obvious by now that the only information being retrieved from that classification is set-membership, which would be identical to that expressed in the cladogram.

Indented-List Classifications (Farris 1976)

The incorporation into classification of certain unique taxa—occasionally Recent but most often fossil—frequently has the effect of creating monotypic taxa of high rank. For example, if a fossil or Recent species were newly discovered and determined to be the sister-taxon of some living higher taxon, under the conventional rules of phylogenetic classification this species must be classified at the same rank as its coordinate sister-taxon. There also has been a tendency to accept the principle of "exhaustive subsidiary taxa": "If a subsidiary level is used within any group it should, as far as possible, be used for all organisms in that group. For example, if a subfamily is used within a family, then the whole family should be divided into subfamilies" (Simpson 1961:18). As Farris (1976:272) points out, this and other rules often have the result that a single species may be placed in its own order, family, subfamily, and genus, thus increasing the number of categorical and taxon names for any one classification. Moreover, Farris (1976:272–73) notes, "several taxa must be constructed, all of which have the same membership—that is the 'several' taxa are nothing more than different names applied to one and the same group."

Whereas most systematists have, in the past, been resigned to accept such a system, its difficulties nonetheless have been generally deplored, and evolutionary systematists have pictured phylogenetic classifications as being particularly vulnerable to these difficulties:

> For the dubious advantage of being able to read off 'phylogeny' from a formal listing of taxa, cladists are willing to pay, as Hull has said, a price too high for many biologists. Species that split off in the Precambrian but give rise to no other species would have to be classed as phyla. Such classifications would be highly monotypic and highly asymmetrical. (Ashlock 1974:97)

Farris has proposed the adoption of indented list classifications as one possible mechanism to classify fossil species. He suggests that monotypic taxa be abandoned except in those circumstances in which a generic name is needed for a species that cannot be assigned to any known genus. According to Farris the only requirements that should be adhered to are "that each monophyletic group must be a taxon, each taxon must be a monophyletic group, and the natural inclusion relationships of the monophyletic groups must be retained by the taxa" (1976:275).

In indented list classifications, ranks are abandoned when indentation within the hierarchy conveys the same information about inclusion relationships. Thus, given the relationships shown in figure 5.18, the list-form classification might be as follows:

CLASS A–F   Species F†
  ORDER A–E
  FAMILY A–B
            Species A
            Species B

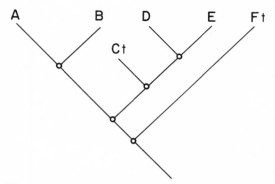

Figure 5.18 A cladogram for six taxa. Daggers indicate fossil taxa. (See text.)

FAMILY C–E
> Species C†
> SUBFAMILY D–E
>> Species D
>> Species E

In this example the fossil species F is placed in the class A–F to denote its inclusion relationships with species A–E, but otherwise F is classified only to genus and species, as required by the rules of nomenclature. Its indentation, in juxtaposition to the order A–E, signifies it as the sister-taxon of the remaining species in the class, however there is no need to classify species F to order or family as this would be redundant. Order A–E is subdivided conventionally except for fossil species C, which is not classified to subfamily but its juxtaposition next to subfamily D–E indicates the two are sister-taxa.

As another example of list-form classification, the vertebrate relationships depicted in figure 5.17 might be represented as follows:

SUBPHYLUM Vertebrata
> Agnatha
> SUPERCLASS Gnathostomata
>> Actinopterygii
>> SUBCLASS Choanata
>>> Amphibia
>>> INFRACLASS Amniota
>>>> Aves
>>>> DIVISION
>>>>> Monotremata
>>>>> COHORT
>>>>>> Metatheria
>>>>>> Eutheria

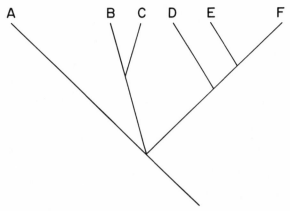

Figure 5.19 A cladogram for six taxa (see text).
(After Farris 1976:276, figure 1.)

In this classification redundant rank and taxon names are elimi-nated. For example, the Agnatha are not classified as to superclass, subclass, infraclass, and so on. Nevertheless, the classification con-veys a precise statement about relationships.

List-form classifications can also be used to depict relationships that are not dichotomous. For the cladogram of figure 5.19, the clas-sification might take the following form:

ORDER A–F
        Species A
      SUBFAMILY B–C
        Species B
        Species C
    FAMILY D–F
        Species D
      SUBFAMILY E–F
        Species E
        Species F

That there is a trichotomy involved is signified by placing spe-cies A at a lower rank than species B–C, and species B and C at a lower rank than the taxon, D–F. Thus, these three groups (A, B–C, and D–F) are equal subtaxa of order A–F, none being a subtaxon of the other.

One of the advantages, then, of indented-list classifications is that redundancy of categorical rank and taxon names is minimized. Also, fossil taxa can be intercalated into or removed from a classification of Recent taxa without disturbing the traditional ranking of the latter. Such classifications may have an important disadvantage in that, until one is thoroughly familiar with how to read them, their interpretation may be confusing, and this could be of some significance when trying to communicate with non-systematists.

The "Plesion" Concept (Patterson and Rosen 1977)

In one of the more extensive discussions about the classification of fossils, Patterson and Rosen have recommended that fossil taxa not be assigned the customary categorical ranks coordinate with their sister-groups. Instead, they suggest that fossil groups be designated "plesions" (in reference to "plesiomorphic sister-group") and inserted into the classification using the convention of phyletic sequencing to specify that each plesion taxon is the sister-group of all taxa listed below it in the classification. In this manner neither subordination nor new categorical and taxon names would be necessary; conventional subordination of ranks would continue to be applied to Recent taxa.

Either single fossil species or fossil groups containing many species may be designated plesions. Furthermore, a given plesion might be coordinate with a Recent taxon assigned to any categorical level. As Patterson and Rosen point out, the use of plesions would be appropriate in general classification of major groups or in more restricted classifications of paleontologically rich groups; plesions would not be needed if the classification were of strictly extant taxa.[10]

The application of the plesion concept can be illustrated with an example taken from Patterson and Rosen's analysis of the relationships and classification of fossil and Recent teleost fish (for an additional example see Wiley 1976:93). Figure 5.20 depicts the hypothesized relationships of a sample of fossil and Recent taxa; Patterson and Rosen (1977:163) propose the following classification:

SUBDIVISION Teleostei
    plesion †*Pholidophorus bechei*
    plesion †*Pholidolepis dorsetensis*

10. There is no reason, of course, why a convention such as "plesion" could not be used for an extant taxon, comprised of one or a few species, that is the sister-taxon of a much larger group.

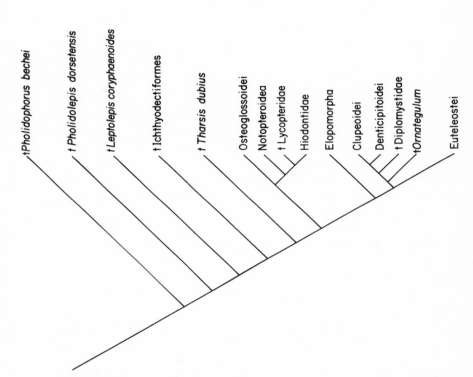

Figure 5.20 A cladistic hypothesis for some Recent and fossil teleost fish (see text). (After Patterson and Rosen 1977:153, figure 54.)

plesion †*Leptolepis coryphaenoides*
plesion †*Ichthyodectiformes*
plesion †*Tharsis dubius*
SUPERCOHORT Osteoglossomorpha
   ORDER Osteoglossiformes
     SUBORDER Osteoglossoidei
     SUBORDER Notopteroidei
      SUPERFAMILY Hiodontoidea
      plesion †Lycopteridae
      FAMILY Hiodontidae
      SUPERFAMILY Notopteroidea
SUPERCOHORT Elopocephala
  COHORT Elopomorpha
  COHORT Clupeocephala
   SUBCOHORT Clupeomorpha
   Order Clupeiformes
     plesion †*Ornategulum*
     plesion †*Diplomystidae*
     SUBORDER Denticipitoidei
     SUBORDER Clupeoidei
   SUBCOHORT Euteleostei

The plesion taxa of this classification include those of a single species and those containing many (e.g., Ichthyodectiformes); they are also seen to be coordinate with supercohort, suborder, and family rank taxa.

The concept of plesions appears to be a useful tool in constructing phylogenetic classifications of large groups of organisms having a rich fossil record. At first thought, the use of phyletic sequencing alone might seem to have the same advantages as the use of plesion, but in order to effect such a classification redundant higher taxon names must be invented. In this example, each of the first five fossil taxa in the classification would have to be placed in a taxon of supercohort rank:

SUBDIVISION Teleostei
  SUPERCOHORT Pholidophorei
    Pholidophorus bechei
  SUPERCOHORT Pholidolepei
    Pholidolepis dorsetensis

SUPERCOHORT Leptolepei
Leptolepis coryphaenoides
SUPERCOHORT Ichthyodectei
Order Ichthyodectiformes
SUPERCOHORT Tharsei
Tharsis dubius
SUPERCOHORT Osteoglossomorpha
SUPERCOHORT Elopocephala

In this case, the creation of new supercohort names is redundant and serves little useful purpose. The plesion concept avoids these difficulties.

Classifying Taxa of Uncertain Status

In routine practice a systematist frequently confronts situations in which the systematic status of taxa is uncertain. This uncertainty may arise from (a) our inability to resolve relationships despite the presence of adequate comparative material, (b) the fact that the taxa may not have been studied in sufficient detail by previous workers, or (c) our inability to resolve relationships because of the absence of adequately preserved material (this is of common occurrence in fossil studies). For the systematist, the problem is how to incorporate such kinds of taxa into a phylogenetic classification.

In cases a and b, the problem almost always involves an unresolved trichotomy in a cladogram analysis. Although a single taxon may be the focus of attention, for example, when an attempt is made to interpolate a fossil taxon into a scheme of relationships based on Recent groups, the uncertainty of the situation extends to some of these Recent taxa as well. Consider the example of Recent taxa, A–C, related as shown in figure 5.21a, and the fossil taxon D, of uncertain relationships. Taxa B and C are determined to have a closer relationship to each other than either has to A, but the precise relationships of D to B and/or C are ambiguous (obviously B, C, and D share some uniting synapomorphy). How is this situation to be expressed in a classification? The simplest answer would be to base the classification on the cladogram (figure 5.21b):

FAMILY A–D
SUBFAMILY A
SUBFAMILY B–D
GENUS B

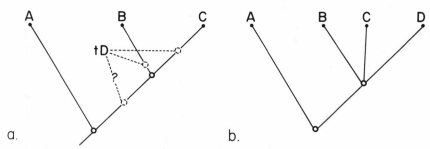

Figure 5.21 A cladistic hypothesis for three Recent taxa and one fossil taxon of uncertain relationships (see text). (After Nelson 1972b:229, figure 2.)

GENUS C
GENUS D

If the classification as a whole is not expressing relationships by phyletic sequencing procedures, then the above method of denoting a trichotomy (and the implied uncertainties in relationships) within the classification will be unambiguous. If, however, portions of the classification include sequencing, then the above notation might be taken to imply that B is the sister-group of C–D. In this case, unresolved trichotomies will have to be tagged for recognition by some convention, perhaps by placing all three groups *incertae sedis:*

FAMILY A–D
  SUBFAMILY A
  SUBFAMILY B–D
    GENUS B *(incertae sedis)*
    GENUS C *(incertae sedis)*
    GENUS D *(incertae sedis)*

Another possibility is that trichotomies might be bracketed within classifications to distinguish them from sequencing:

FAMILY A–D
  SUBFAMILY A
  SUBFAMILY B–D
    ⌈GENUS B
     GENUS C
    ⌊GENUS D

Patterson and Rosen (1977), in dealing with the uncertainties discussed here, adopted the following conventions: (1) fossil taxa were tagged plesions and their relationships were then expressed using phyletic sequencing; (2) Recent taxa were subordinated in the traditional manner; (3) unresolved trichotomies within Recent taxa were expressed by a simple listing (at the same rank) of the three taxa; and (4) the use of *incertae sedis* was restricted to fossil groups whose relationships could not be resolved. Most classifications of major groups will sometimes include taxa that are known to be non-monophyletic. Such taxa may not have been the focus of investigation but need to be included in the classification for one reason or another, or for various reasons they have not been the subject of detailed analysis by previous workers. Patterson and Rosen suggest that such taxa be placed in quotation marks within classifications to distinguish them from taxa assumed to be monophyletic.

Conclusions

The important conclusions of this chapter on classification are as follows:

1. The history of systematic biology documents repeated examples of attempts to classify organisms into A and not-A groups, that is, those groups defined by derived attributes and those defined by primitive characteristics. That history also documents a tendency on the part of systematists to recognize only A groups as being natural and to eliminate not-A groups from classifications. A theme of the chapter is that phylogenetic classification provides a solution to the problem of obtaining natural classification, in contrast to the proposals of other systematic philosophies, because the method of phylogenetic systematics is designed to discover and classify A groups.

2. Linnaean classification systems consist of lists of taxa hierarchically arranged. The only information inherent in a Linnaean hierarchy is set-membership.

3. Branching diagrams and Linnaean classifications have the same logical structure, that is, of nested sets of taxa. Because of this, Linnaean classifications must inevitably be based on some type of branching diagram.

4. There are three main kinds of branching diagrams: phenograms, evolutionary trees, and cladograms. Phenograms have not generally been favored as a basis for general classification systems

because they lack an obvious connection with the expected hierarchical structure of nature produced by the evolutionary process. Evolutionary trees and cladograms, on the other hand, have such a connection.

5. The information content of evolutionary trees cannot be expressed readily in Linnaean classifications, because specified ancestors cannot be easily incorporated into classifications. Supraspecific ancestors, in any case, do not exist in nature.

A cladogram can express the underlying set-membership of an evolutionary tree and thus can serve as a basis for classification. The logical structure of cladograms and Linnaean hierarchies is identical, and thus the set-membership of the two can be made isomorphic.

6. Evolutionary classifications, as advocated by Mayr, Simpson, and others, cannot be recommended because the logical structure of the Linnaean hierarchy makes it impossible to express simultaneously both genealogical relationships and a measure of overall similarity. Consequently, the classificatory procedures of evolutionary systematists make it difficult to retrieve from their classifications any of the information putatively used to construct the classification.

7. We recommend that phylogenetic classifications based on cladograms be adopted as general reference systems in biology. Such phylogenetic classifications explicitly avoid acceptance of not-A sets. They have been shown to predict character distributions, including general phenetic resemblance, more efficiently than phenograms or "evolutionary" classifications. They are also isomorphic with the hypothesized genealogical relationships of the included taxa. Genealogy determines set-membership, thus genealogical information (hypotheses) can be incorporated directly into a classification. Using conventions such as subordination, phyletic sequencing, the indented-list procedure of Farris (1976), and the plesion concept of Patterson and Rosen (1977), it is possible to construct phylogenetic classifications for any group of organisms, no matter how taxonomically diverse it might be. Phylogenetic classifications, because they attempt to express the underlying pattern of nature produced by evolution, yield natural classifications and thus provide the soundest foundation for future studies in comparative systematic biology.

Chapter

6

Systematics and the Evolutionary Process

THE FIRST part of this book has been devoted to biological systematics: method and theory pertaining to the elucidation of patterns of phylogenetic relationship among organisms. In so doing, we have adopted the view that for basic phylogenetic research, all that is required is the assumption that life has evolved, i.e., that all life constitutes a monophyletic assemblage. The formulation and testing (including eventual corroboration) of such hypotheses of phylogenetic relationship are of great intrinsic scientific interest. Highly corroborated hypotheses of phylogenetic relationship are a fundamental goal of research in biological systematics and need no further rationale.

Yet there clearly are further uses for such hypotheses. We have already considered one—classification—in the preceding chapter. Classifications express the hierarchical arrangement of taxa and give names to monophyletic groups, thereby recognizing the status of such groups as entities and allowing biologists to talk about them conveniently. The phylogenetic system expressed in cladograms and classifications is the reservoir of corroborated hypotheses on the history of life, which forms the basis for all further predictions and generalizations about the nature of life. The importance of this undertaking, and the complexity and enormity of the task, is sufficient to occupy the lifetime attention of a professional systematist working on only a fraction of the earth's past and present organic diversity.

Yet, clearly, there is even more to the implications of the phylogenetic system than recognizing and naming monophyletic groups. The phylogenetic system codifies and epitomizes all that we think we know about the genealogical history of life on earth. And of course, in this context, "genealogy" is just another expression for "evolution." If we need know nothing specific about how life has evolved to formulate and test genealogical hypotheses, there is still the opposite side of the coin to consider: can corroborated genealogical hypotheses be used in any way to study the process of evolution? And if so, how? In the remainder of this book, we shall develop the theme that such corroborated phylogenetic hypotheses can indeed be used to investigate the evolutionary process, and that phylogenetic hypotheses of the kind treated in the first half of this book are consistent with, and suggest further, lines of inquiry and kinds of hypotheses of evolutionary process somewhat different from those of the standard so-called "synthetic" paradigm.

Relationship between Evolutionary History and Evolutionary Process

The general form of what we take to be the correct relationship between phylogenetic history and the scientific study of the evolutionary process is simply stated: hypotheses of evolutionary mechanisms (derived *de novo* or taken in whole or in part from other theories) generate predictions about results (historical patterns) which are then compared with highly corroborated hypotheses of phylogenetic and distributional patterns. The predictions are either verified or found not to agree with the corroborated hypotheses of actual events. Thus, in its simplest form, the study of the evolutionary process is no different from any other subject of scientific enquiry.

The conclusion that evolution can be studied scientifically will, of course, come as no surprise to evolutionary biologists. But it is necessary to make this simple point because evolutionary theory, since its inception as a subject worthy of serious scientific consideration, has been plagued by an assumption-ridden inductive process whereby historical patterns are examined, narratives dreamt up to explain them (in itself a valid way of formulating new ideas), and further cases made to fit, rather than used to assess, prior assumptions and notions of process. Such a procedure constitutes a misuse of historical patterns and has impeded progress in at least some areas of evolutionary theory. We develop a more detailed critique of some aspects of contemporary evolutionary theory below. For now, just to provide a single example of the misuse of historical patterns in the analysis of evolutionary mechanisms, consider the way in which the concept of "natural selection" is used to explain why an organism has a certain morphological configuration. There may or may not be scientific grounds for preferring a neo-Lamarckian, neo-Darwinian, or even a neo-Kiplingian version of "why the giraffe got its long neck," but the fact remains that the results of history (i.e., existence of a monophyletic assemblage of artiodactyls, Family Giraffidae, with, among other things, long necks) are often explained as "just so" stories, rather than being used to provide the critical evidence for the evaluation of the relative merits of conflicting views on the nature of the evolutionary process itself.

On the other hand, it is equally obvious that historical patterns— or some concept of historical events—have been absolutely essential to the development and progress of evolutionary theory, and this holds true for each of the several levels of evolutionary phenomena discussed in this chapter. Microevolutionary phenomena (changes in gene content and frequency, and concomitant effects within populations and species) are perhaps the best understood of all levels of the evolutionary process. This is true simply because, at this level, the phenomena are directly susceptible to experimental analysis in the laboratory, as well as to independent theoretical mathematical analysis and field investigation. Thus hypotheses are tested directly, empirically. Less well appreciated is the fact that the data derived in, say, a routine *Drosophila* population-cage experiment are highly corroborated hypotheses (to the point of being fact, since all individuals are known to be descendants of at least some of the individuals from the initial population) concerning the historical relationships of generations. These corroborated hypotheses provide the historical background against which the nature of the observed changes in gene frequency (if any) are evaluated. This is, of course, a trivial example, as these historical hypotheses are the natural outcome of rigidly controlled laboratory conditions. Yet it is the historical nature of the data that renders such experiments of any relevance whatever to evolutionary theory. At higher phenomenological levels, the historical data are more explicitly acknowledged as important, but become more susceptible to doubt, because they are hypotheses· of varying degrees of corroboration. We explore the kinds of historical hypotheses appropriate to the various levels of the evolutionary process in depth later in this chapter. In essence, the use of historical hypotheses as criteria for evaluating predictions from theories of process (and thus the theories themselves) is the same at all levels of evolutionary analysis.

Contemporary Evolutionary Theory: A Basic Characterization

In the earlier chapters dealing with speciation and its relationship to systematics, we characterized the essential aspects of various particular theories of speciation, and related these theories to definite geometric concepts of the results of these processes. Each speciation theory leads to its own set of expected historical outcomes, each with its own geometric configuration represented on branching diagrams—specifically, phylogenetic trees. We contrasted two particular modes of speciation: phyletic transformation of a lineage, resulting in the derivation of a descendant species directly from its ancestor, as opposed to concepts of lineage splitting, where genetic isolation occurs to separate descendant species from their ancestors. We briefly noted that in the former case (transformation within a lineage), new species arise as a by-product of the transformation of the genetic properties of a lineage, and thus are viewed largely as arbitrarily delineated segments of that lineage. This transformational view of species is actually in conflict with the definition of species adopted earlier in the book, inasmuch as the pattern of parental ancestry and descent defining a species is very much present, by definition, throughout the life-span of the entire lineage. Therefore, under our adopted definition of "species," [and as Wiley (1978) has cogently pointed out] a continuous lineage, unbroken by a splitting event where some fraction of the lineage is split off and becomes a new species, must be considered a single species no matter how long it persists or how much phenotypic change occurs during this time span. We agree with Wiley that this concept of "species" agrees exactly with Simpson's (1961) concept of an evolutionary species, while remaining consistent with Mayr's (1942), as we noted earlier (chapter 3, p. 93).

In contrast to the transformational concept, our adopted definition of species requires that they have a definite origin (from an ancestral species), and, logically, a definite termination. This is to say that species are discrete entities. All theories of speciation involving splitting mechanisms (briefly reviewed in chapter 4) implicitly take the position that species are ontologically "real" entities in nature. This view has been further developed from a philosophical point of view (for slightly different reasons) by both Ghiselin (e.g., 1974) and Hull (e.g., 1976). We reiterate these points at this juncture because it is evident that these two contrasting points of view concerning species and their origins can be generalized even further to characterize two prevailing and conflicting approaches to evolutionary theory (especially, but certainly not exclusively, within the synthetic paradigm). Furthermore there is a general geometric scheme for the results of the evolutionary process that conforms to each of these views. Distinguishing between these two approaches clarifies the basic issues in evolutionary biology and points the way towards some alternative approaches to some of the older problems in evolutionary theory.

This dichotomy in evolutionary theory is simply stated. It can best be characterized by the two contrasting definitions of evolution prevalent today, which highlight a duality in what evolutionists take to be the central problem in evolution. One view holds that evolution is primarily a phenomenon of the transformation of genes, their frequencies, and their products (reflected in the phenotype, behavior, and physiology). All modern definitions of evolution include this transformation aspect, and most restrict themselves to it:

> Organic evolution is a series of partial or complete and irreversible transformations of the genetic composition of populations, based principally upon altered interactions with their environment. It consists chiefly of adaptive radiations into new environments, adjustments to environmental changes that take place in a particular habitat, and the origin of new ways for exploiting existing habitats. These adaptive changes occasionally give rise to greater complexity of developmental pattern, of physiological reactions, and of interactions between populations and their environment. (Dobzhansky et al. 1977)

One of the many implications of this approach, as codified and symbolized by definitions of this sort, is the particular point developed earlier: that species emerge as a logical consequence of such transformations within lineages. It is, of course, possible to retain this general concept of evolution and to discuss the origin of new, discrete species via genetic isolation. But the particular notion that species arise as a consequence solely of accumulated changes within lineages is the view most consistent with the general transformational concept and, more importantly, has historically dominated the thinking of most paleontologists and not a few geneticists.

For most contemporary biologists, the transformational approach is generally developed in terms of adaptation, usually via natural selection. In this view, the central problem in evolution is the transformation of the genotype which occurs as organisms adapt in response to selection pressures. Adaptation is the ultimate *raison d'être* for change in gene frequency. Natural selection is the effective agent. Mutations ultimately, and various mechanisms (such as recombination) proximally, supply the requisite panoply of variation upon which selection acts. This is the core of the neo-Darwinian version of the transformational doctrine.[1] Our purpose is not to attack the neo-Darwinian catechism itself—indeed, we find it is probably as highly corroborated as any other equally complex set of propositions in biology—but rather to show that an exclusive and rigid adherence to the basic transformational view has resulted in the inappropriate application of these concepts to all aspects of evolutionary phenomena. Thus adaptation via selection underlies most previous attempts to explain the results of the evolutionary process. We have already commented that such adaptational scenarios are themselves not appropriate to the scientific study of evolution (either the process or its results). Considered as a research strategy, an exclusively transformational approach to major problems of rates and modes of evolu-

---

1. We use the term "neo-Darwinism" to refer to the body of theory that identifies mutation as the ultimate source of genetic variation within populations and states that population sizes are limited and that there is therefore a resultant pattern of differential reproduction (natural selection) within populations which systematically alters gene frequencies since only some of the variation present in one generation can be present in the next. "Syntheticism" refers to a loosely defined body of work carried out from the 1920s to the early 1950s, marked especially by a tendency to extrapolate neo-Darwinian principles to encompass speciational and interspecific evolutionary phenomena. Thus the terms are not synonyms, as Simpson (1978) has recently pointed out. To the extent that we accept selection as the best available deterministic hypothesis accounting for change in gene content and frequency within populations, we remain neo-Darwinians. The failure of the syntheticists to incorporate speciation theory into a testable theory of macroevolution leads us to avoid the term in describing the views presented here.

tion has led to the development of a set of propositions which seem, at base, rather difficult to test. In short, the within-population phenomena of mutation, variation, and selection have been extrapolated across the board to embrace all aspects of evolution, including so-called macroevolution. In so doing, an important ingredient is usually missing, an ingredient that makes all the difference in the nature and composition of theories of the evolutionary process.

That missing ingredient is speciation. Alongside the dominant theories of transformation has been another theme, that of the evolution of taxa. Although one rarely finds in the literature definitions of evolution conforming to this alternate approach, it is clear that many biologists (historically, mostly neontological systematists) who have studied species and theorized on speciation implicitly adopt the view that evolution is quintessentially a matter of the origin of new taxa—i.e., species. This, the "taxic" approach to evolution, simply alleges that, at its core, it is species that evolve—not individuals, and still less, anatomical or genetical properties of individuals. In the process of speciation, such genetic change might well occur, even as an overwhelming expectation. In an earlier characterization of the differences between these two approaches to evolutionary theory, Eldredge (1979b) argued that the taxic view is the more general, embracing concept of the two, in that speciation cannot be understood strictly with reference to accumulated changes in genes, their frequencies, and their products, but that speciation provides the meaningful context for understanding and interpreting any such changes in genotype, whether adaptive or not. We do not pursue this view here. It is clear that, under the concept of species adopted in this book, that a great deal of genetic and morphological change can, at least theoretically, occur within the total time of existence of a single species, without any splitting events occurring. To define evolution simply as the origin of new species would be to deny that phyletic change within a lineage (single species) is also evolutionary. Rather, the gist of the problem is that the existence of discrete species—and the attendant problems of their origins, subsequent histories, and eventual demises—represents a phenomenological level distinct from that of the population.

One of the logical consequences of the foregoing observation has been badly neglected: considerations of all macroevolutionary phenomena (defined for the moment as evolutionary phenomena pertaining to genera and taxa of higher categorical rank) cannot be addressed without explicit reference to species—their origins, histories, and patterns and modes of extinction. As we shall now show, *nearly all attempts by syntheticists to deal with macroevolution, at least until recently, fail even to mention species, much less to incorporate hypotheses pertaining to species into the analysis.* In short, the transformational approach has given us a view of evolution based on the assumption that the entire tree of life is one huge, smoothly continuous gene pool. In this view, evolution is merely a matter of progressive change of intrinsic features and its products are appropriately to be understood almost strictly in terms of adaptation and selection (although genetic drift enters the picture as a random component in some formulations). Discrete species become an embarassment instead of the very key (as we believe) to deriving more accurate and more testable hypotheses pertaining to evolutionary phenomena at and above the species level.

Contemporary Macroevolutionary Theory: Issues and Transformational Explanations

Any summary and critical evaluation of a body of theory must have some limits, however loosely defined. In this case, we construe "contemporary" to refer to work performed during the past fifty years or so, especially the so-called synthetic theory. It is our belief that the majority of Western biologists interested in macroevolution still fundamentally adhere to many of the basic tenets outlined in the 1930s, 1940s, and 1950s. This is not to say that theory developed even earlier is uniformly uninteresting or without relevance, but merely that the more common notions held today were formulated and codified during the past half-century or so, and were themselves partly based on, and partly a reaction against, earlier work. Nor do we suggest that authors not explicitly numbering themselves among the syntheticists (or, indeed, explicitly dissociating themselves from the syntheticists, or being disavowed by the latter) have had no important contributions to make. But there is no discrete "school" (even among the "saltationists") that can be easily labeled and characterized. The isolated writings of Willis (1940), Goldschmidt (1940), Schindewolf (e.g., 1950), Cuénot (1932), Rosa (1931), Grassé (1973), and Løvtrup (1977), to name a few biologists who have addressed evolutionary theory in more or less nonsyntheticist terms, are a most heterogeneous array, united only by the fact that each deviates in one or more ways from orthodox syntheticism, and all display a general tendency to adopt a transformational viewpoint. In focusing on the synthetic school (itself rather heterogeneous when examined closely), not only do we emphasize the most widely held views, but we are also able to discern trends and styles of thought which are also present in the writings of most of the "nonsyntheticists" as well. The point of this review of contemporary macroevolutionary theory is to pick out such basic modes of thought in addition to exposing the core of the theory itself, in the belief that analysis of the epistemological structure of such existing theory is essential to the development of a protocol for future scientific research in evolutionary theory.

Simpson (1944) spoke of "tempo" and "mode" as a useful dichotomy of evolutionary phenomena. Tempos, of course, refer to evolutionary rates: genetic, morphological, and taxonomic, whereas modes refer to styles of change (for Simpson 1944, there were three distinct modes: speciation, phyletic evolution, and "quantum evolution"). Specific topics usually considered as macroevolutionary include all types of rates viewed as a problem among species and taxa of higher categorical rank, and, especially, the very nature and mode of origin of taxa of higher categorical rank.

That species are real, i.e., that species have ontological existence in nature, and display beginnings, continued history, and discrete ends, is one of the more important premises of this book. That taxa of categorical rank higher than species do not exist in precisely the same sense as do species is crucial both to an appraisal of past and current macroevolutionary theory, and to our own elaboration of such theory. What all taxa, from species up through kingdoms, do share is presumed descent from a single ancestral species. What they do not share are similar reproductive patterns. During the entire time interval of their existence, species display patterns of parental ancestry and descent within themselves, but not without, as an overwhelming generalization. It is these units (species) that evolve, largely by producing isolated descendant units of like kind (more species). Taxa of higher categorical rank cannot be defined with reference to reproductive patterns (a within-species phenomenon) except for the trivial observation that the same disruption of parental ancestry and descent among species by definition applies to genera, families, and so on up the Linnaean hierarchy. Thus the definition of taxa above the rank of species is identical for all ranks: a monophyletic group comprised of one or more species. Such taxa, ideally, at least, reflect our knowledge of life's genealogical history. As discussed in chapter 2, these groups of species can be recognized in a fairly objective and scientific manner. As hypotheses of monophyletic pathways of descent, they surely have relevance to our elaboration, and testing, of hypotheses bearing on the evolutionary process. But they do not evolve in the same sense as species, simply because they are not the same kinds of biological entities. They are all collections of species. The evolution of a genus, an order, or a kingdom is fundamentally the same thing, amounting to varying patterns of origin and survival of component species.

The distinction between the nature of species, on the one hand, and taxa of higher categorical rank on the other, is crucial, simply because much has been written on the subject of the origin of taxa of higher categorical rank. And much of this literature is written under the tacit assumption that such taxa possess an ontological status over and above monophyly of descent. For example, Simpson (e.g., 1953:340 ff. and elsewhere) clearly recognizes the fact that taxa of higher rank are recognized *ex post facto* and thus are constructs of the systematist's mind, not existing in nature in any real sense; yet he has recently written (Simpson 1978:77) that " 'Species selection' *is simply natural selection at a particular taxonomic level,* but Stanley has emphasized well that major changes are best studied at this (*or higher*) levels" (italics added). This statement directly imputes an ontological equivalence in status to species and taxa of higher rank—the very confusion we believe has beclouded much of macroevolutionary theory to date.

If taxa of rank higher than species do not exist, or evolve, except as their component species do, we must seriously ask: what is the problem? Is there any sense to speak of macroevolution at all? In a series of recent papers, Gould, Raup, Schopf, and other colleagues have investigated shapes of "clades"—i.e., patterns of relative and absolute species diversity within both "real" (supposedly) monophyletic taxa (using data drawn from the systematic literature) and computer-simulated taxa. Their work examines the extent to which patterns of change in diversity within monophyletic groups reflect deterministic (non-random) processes versus "random" processes. (See p. 298 for a more complete discussion of the meaning of these terms and the significance of this research.) There appears to be a significant nonrandom component to such changes in diversity patterns, which alone suffices to show that a macroevolutionary theory must have a strong deterministic component. Although we will be critical of contemporary macroevolutionary theory in its insistence on reductionist transformational explanations, nevertheless many of the issues raised retain their importance and can, we believe, be approached if properly restated.

The Adaptive Landscape

In 1932, Sewall Wright introduced the concept of the adaptive landscape. In its original formulation, the topographic grid of hills and valleys served as a graphic, two-dimensional depiction of the relative "adaptive value" of various "harmonious combinations" of a number of loci, each with a number of alleles. Dobzhansky (1937 and later editions; see also 1970 and Dobzhansky et al. 1977) has, perhaps more than any other author, utilized this graphic representation to pinpoint some of the more crucial issues in evolutionary theory. Dobzhansky and his coauthors point out that, with the two-dimensional landscape, the two axes "symbolize the allelic variants of two gene loci. The Darwinian fitness (adaptive value) of the combinations of these alleles is symbolized by contours, like a topographic map" (1977:168). Hence, they retain the force and gist of Wright's original metaphor, although the equation of fitness (defined elsewhere in the book as differential reproduction) with individual vigor or survival is problematical. However, Wright himself (1932) explicitly wrote in terms of relative survival value of allelic combinations. At its core, the concept of the adaptive landscape allows one (but not necessarily Wright himself—see below) to picture natural selection as directly favoring the "more harmonious combinations" (peaks) over the less favorable allelic combinations, given the full range of such possibilities.

The problem immediately arises: how does one (metaphorically speaking) get from one peak to another? Or, as Wright (1932:358) put it: "The problem of evolution as I see it is that of a mechanism by which the species may continually find its way from lower to higher peaks in such a field." Wright went on to elaborate a hypothesis whereby "the course of evolution through the general field is not controlled by direction of mutation and not directly by selection, except as conditions change, but by a trial and error mechanism consisting of a largely nonadaptive differentiation of local races (due to inbreeding balanced by occasional crossbreeding) and a determination of long term trend by intergroup selection" (1932:365). The important point here is that Wright explicitly intended the "adaptive landscape" to apply strictly and solely to various allelic combinations *within species* (especially within semi-isolated subpopulations of species; see Wright 1931, 1945:415 and many of his other works). Above that level, some other process of intergroup selection not involving an adaptive-landscape motif was envisioned to be in operation. Further, it is of interest that Wright emphasized a "trial and error" mechanism, rather than either mutation or selection, as the major means of travel between peaks. Later workers, most notably Dobzhansky (1937) and Simpson (especially 1953, after abandoning the "inadaptive phase" of quantum evolution as conceived in 1944) agreed about the (non)role of mutation in this matter, but came down firmly in favor of a model of selection as a means, not just of climbing peaks, but of traversing the valleys as well. We shall not pursue the issue here, since it is a within-population and within-species problem.

Or, at least, the issue was at the within-species level when originally formulated and, in part, proselytized in Dobzhansky's several books. Through the three editions of his *Genetics and the Origin of Species,* Dobzhansky (1937, 1941, 1951) progressively augments his discussion and utilization of Wright's imagery. By 1951, Dobzhansky had explicitly changed Wright's original context, applying the concept of the adaptive landscape to gene combinations among species, families and on up the Linnaean hierarchy:

> The enormous diversity of organisms may be envisaged as correlated with the immense variety of environments and of ecological niches which exist on earth. But the variety of ecological niches is not only immense, it is also discontinuous. One species of insect may feed on, for example, oak leaves, and another on pine needles; an insect that would require food intermediate between oak and pine would proba-

bly starve to death. Hence, the living world is not a formless mass of randomly combining genes and traits, but a great array of families of related gene combinations, which are clustered on a large but finite number of adaptive peaks. Each living species may be thought of as occupying one of the available peaks in the field of gene combinations. The adaptive valleys are deserted and empty. (Dobzhansky 1951:9 ff.)

Dobzhansky then goes on, in the following paragraph, to extend this generalization up the Linnaean hierarchy, concluding that: "The hierarchic nature of the biologic classification reflects the objectively ascertainable discontinuity of adaptive niches, in other words the discontinuity of ways and means by which organisms that inhabit the world derive their livelihood from the environment." Thus is the within-species topographic conception of Wright translated into a model for the diversity of life. The interesting point about these rather beguiling quotations from Dobzhansky is that, just as Wright saw for his conception, they imply an immediate problem: the central problem in evolution is, how does a *taxon* (i.e., not a population with a trial and error means of arriving at relatively more harmonious allele combinations) get from one peak to another? How are valleys traversed and hills climbed? In one neat and logical stroke, Dobzhansky equated a within-species problem with the entire problem of evolution and, more subtly, suggested that the answer to the one was the answer to the other. The key word, in any case, remains the same: "adaptive." Wright's "trial and error" of allelic combinations was tacitly dropped, and the way was cleared for a true synthesis of the data of systematics (including paleontology) with the theory and results of experimental and mathematical population genetics. If the concept of the adaptive landscape neither completely accounts for, nor covers all topics addressed by "syntheticist" writings on macroevolution, it at least clearly symbolizes the route (and it *is* a logical one) that writers have overwhelmingly taken to bridge the perceived gap between microevolution within populations and species, on the one hand, and the phenomena of evolution "above the species level," as Rensch (1960) puts it, on the other hand. The conclusion that all evolutionary phenomena were not only consistent with genetic theory, but could actually be explained by reference to such "first principles," was obviously an attractive one. Our purpose in this chapter is not to say that such findings (e.g., in genetics) are in fact inconsistent with the data of systematics, but rather that the data and theory of genetics and systematics have been integrated in an inappropriate fashion. And the reason for this is that a model, especially but not solely Wright's adaptive landscape, which was developed for a multiple loci, multiple allelic circumstance within species, is inappropriate for wholesale translation to among-species levels—let alone to higher taxonomic levels.

Nearly all major works pertaining to modern, neo-Darwinian formulations of evolutionary theory address the problems of macroevolution to some extent. However, we agree with Mayr (1963:586) that the major works on what he terms "transspecific" evolution are those of Simpson (see especially, but by no means exclusively, 1944 and 1953) and Rensch (1960, an English version of a work first published in German in 1947). An examination of their works is therefore in order, along with ancillary writings by these and other authors.

Rensch (e.g., 1960:1, 57, and 97) repeatedly asks the rhetorical question: are the processes accounting for formation of races and species (mutation, selection, etc.) sufficient to explain transspecific evolution "leading to the emergence of new genera, families, orders, classes etc., and hence to the formation of new organs and new types of organizations" (p. 57)? He concludes that, whereas there are rules which govern transspecific evolution (mostly environmental and "somatic" constraints on the possibilities for anatomical modification), we basically must say "yes." In each of the summations of his lengthy considerations of rates (p. 96) and cladogenesis (p. 279) occur statements such as the following (p. 279): "Summing up, we may state that the evolution of new structural types and of new organs needs no other explanation than specific and generic differentiation, i.e., the combined effect of mutation and selection." Thus it is quite clear that Rensch (1) views evolution fundamentally as a problem of the transformation of anatomical features and (2) denies the existence of different phenomenological levels in evolution. The processes which Rensch discusses (e.g., alteration of developmental pathways) are (as in the case of Goldschmidt—see footnote 4 to this chapter) undoubtedly relevant to a complete theory of macroevolution, but only if considered in the proper context, i.e. with regard to the phenomenological levels developed later in this chapter.

Simpson (1944) set out with the avowed purpose of achieving a "synthesis of paleontology and genetics." In a sense, he succeeded admirably, and no longer was there any rational doubt that the principles of genetics were somehow connected to, and consistent with, the results of the evolutionary process, bits and pieces of which are to be found in the fossil record. He also achieved his goal of suggesting "new ways of looking at facts and new sorts of facts to look for" (1944:xviii) in the work.

The intellectual tradition in vertebrate comparative anatomy and paleontology vis à vis notions of organic evolution began amid a strong revulsion which set the tone for nearly all subsequent discussions. Cuvier and Agassiz, both comparative anatomists and paleontologists, were outspoken in their opposition to any notion of evolution. The stumbling block that anatomists have cited over and over again is the problem of transforming one complex structure into another. They have been, in short, overawed by the complexity and supposed magnitude of differences between fully developed structures. They have worried about obviously inviable, but supposedly necessary, "intermediate" structures, such as useless rudimentary eyes, *in statu nascendi*.[2] Proevolutionist paleontological writing accordingly has been geared to counteract these objections, and Simpson's works (see especially 1944, 1953) are certainly no exception. Simpson (1944:89) adopted Wright's adaptive landscape and extended it to embrace among-species phenomena before Dobzhansky did, alleging that it portrays "the relationship between selection, structure, and adaptation." Inasmuch as Wright (1932) (a) deliberately downplayed the direct role of selection, (b) wrote of multiallelic loci, not anatomical structures, and (c) used the expression "adaptive value" (i.e., survival value of particular loci) and not "adaptation," it is immediately evident that Simpson at the very outset modified Wright's model for some larger purpose. The remainder of Simpson's characterization of this landscape is quoted in full:

> The field of possible structural variation is pictured as a landscape with hills and valleys, and the extent and directions of variation in a population can be represented by outlining an area and a shape on the field. Each elevation represents some particular adaptive optimum for the characters and groups under consideration, sharper and higher or broader and lower, according as the adaptation is more or less spe-

2. W. J. Bryan, as recently as 1925, raised similar objections during the Scopes trial, and the theme persists in some creationist literature to the present day.

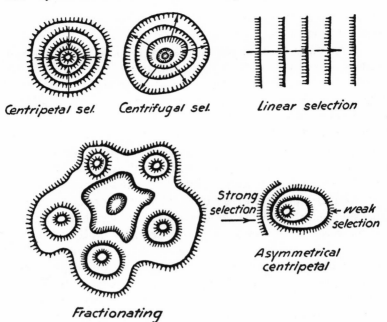

Figure 6.1 Simpson's depiction of the adaptive landscape. His original caption follows: "Selection landscapes. Contours analogous to those of topographic maps, with hachures placed on downhill side. Direction of selection is uphill, and intensity is proportional to slope." (From Simpson 1944:90.)

cific. The direction of positive selection is uphill, of negative selection downhill, and its intensity is proportional to the gradient. The surface may be represented in two dimensions by using contour lines as in topographic maps (figures 11, 12) [reproduced here as figures 6.1 and 6.2]. The model of centripetal selection is a symmetrical, pointed peak and of centrifugal selection, a complementary negative feature, a basin. Positions on uniform slopes or dip-surfaces have purely linear selection. The whole landscape is a complex of the three elements, none in entirely pure form. To complete the representation of nature, all these elements must be pictured as in almost constant motion—rising, falling, merging, separating, and moving laterally, at times more like a choppy sea than like a static landscape—but the motion is slow and might, after all, be compared with a landscape that is being eroded, rejuvenated, and so forth, rather than with a fluid surface. (Simpson 1944:89–90; repeated nearly verbatim 1953:155–56)

There follow immediately (Simpson 1944:90 ff.; 1953:157 ff.) several pages wherein this version of the landscape metaphor is ap-

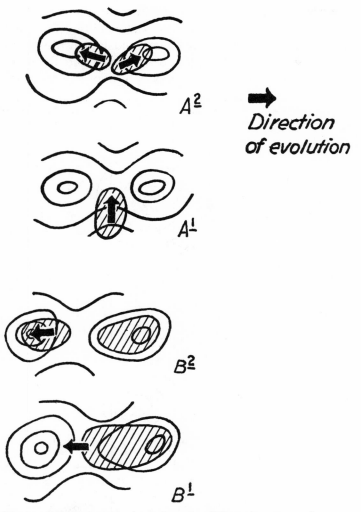

Figure 6.2 Simpson's depiction of splitting phenomena in terms of the adaptive landscape. His original caption follows: "Two patterns of phyletic dichotomy; shown on selection contours like those of [figure 6.1]. Shaded areas represent evolving populations. A, dichotomy with population advancing and splitting to occupy two different adaptive peaks, both branches progressive; B, dichotomy with marginal, preadaptive variants of ancestral population moving away to occupy adjacent adaptive peak, ancestral group conservative, continuing on same peak, descendant branch progressive." (From Simpson 1944:91.)

plied directly to equid evolution. It is a superb example of what we have earlier termed a "scenario" (see also Tattersall and Eldredge 1977). We here reproduce Simpson's original illustration presented in conjunction with his discussion (figure 6.3). In this figure there are three subfamilies (two of which are nonmonophyletic; see MacFad-

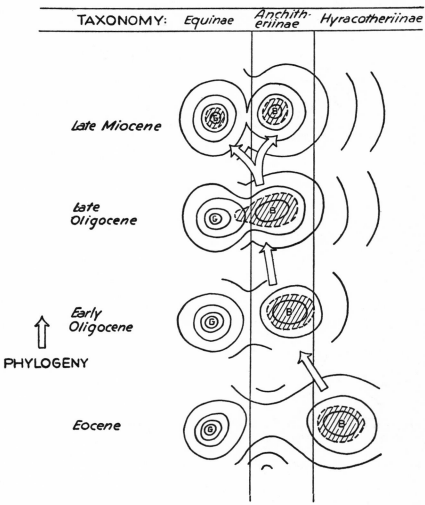

Figure 6.3 Simpson's depiction of "major features of equid phylogeny and taxonomy represented as the movement of populations on a dynamic selection landscape." Simpson referred the reader to the section of text quoted in full here for additional explication. (From Simpson 1944:92.)

den 1976, for a recent discussion of equid phylogeny), whose phylogenetic trees and adaptations are simultaneously depicted as a series of four successive adaptive topographies. The point to be emphasized here is that the discussion focuses on the shift from one peak to another, or the old dilemma of the transformation of one morphological configuration into another. The illustration (our figure 6.3) is particularly concerned with the shift from browsing to grazing (browsers later becoming extinct) which, as Simpson notes, involves morphological features used to characterize and define the three subfamilies listed on the diagram. The critical part of Simpson's discussion is quoted in full, as evidence of our characterization:

> In the Eocene browsing and grazing represented for the Equidae two well-separated peaks, but only the browsing peak was occupied by members of this family. That peak had moderate centripetal selection, which was asymmetrical, because one kind of variation, on one side, away from the direction of the grazing peak (teeth lower than optimum, and so forth) was more strongly selected against than on the other side, in the direction of the grazing peak.
>
> As the animals became larger—throughout the Oligocene, especially—the browsing peak moved towards the grazing peak, because some of the secondary adaptations to large size (such as higher crowns, as previously discussed) were incidentally in the direction of grazing adaptation. Although continuously well adapted in modal type, the population varied farther toward grazing than away from it, because of the asymmetry of the browsing peak. In about the late Oligocene and early Miocene the two peaks were close enough and this asymmetrical variation was great enough so that some of the variant animals were on the saddle between the two peaks. These animals were relatively ill-adapted and subject to centrifugal selection in two directions. Those that gained the slope leading to grazing were, with relative suddenness, subjected to strong selection away from browsing. This slope is steeper than those of the browsing peak, and the grazing peak is higher (involves greater and more specific, less easily reversible or branching specialization to a particular mode of life). A segment of the population broke away, structurally, under this selection pressure, climbed the grazing peak with relative rapidity during the Miocene, and by the end of the Miocene occupied its summit. Variants on the browsing slope tended by slight, but in the long run effective, selection pressure to be forced back onto that peak, and the competition on both sides from the two well-adapted groups caused the intermediate, relatively inadaptive animals on the saddle to become extinct. Thereafter browsing and grazing populations were quite distinct, each differentiated in minor ways. The browsing types

eventually, at about the end of the Tertiary, failed to become adapted to other shifts in environment and became extinct, while the grazing types persist today. (Simpson 1944:91–93; 1953:158–59)

The essence of this passage is clear: the evolution of taxa of rank higher than species is to be explained by reference to the modification of those anatomical structures on which the recognition and definition of those taxa are based. Such changes are understood to be purely the result of the action of various modes of natural selection acting on a groundmass of phenotypic variation (itself a reflection of underlying genetic variation). Apart from the manifest difficulty one would have in testing any of the specific statements about selection vis-à-vis equid fossils, the statement is a logical extrapolation of what was then, and remains today, generally understood to be the nature of the dynamics of within-population genetic change. We merely wish to point out two things about this passage: (1) it represents the very essence of Simpson's approach to macroevolution— an approach which indeed does blend paleontology with genetics and is also consistent with at least 100 years of previous paleontological thought on the central problem of evolution (transformation of structure), and, more importantly, (2) nowhere in the discussion are species mentioned. Unlike Wright's admittedly terse comments on the subject (see page 273), there is no recognition of the possibility that species exist and are to be regarded as units of evolution and as taxa distinct in kind from taxa of higher rank (elsewhere and in different contexts Simpson acknowledges much of this to be the case; see also page 196). Simpson also fails to point out that at the among-species level, the actual geometry of evolution might be radically different, such that the bald extrapolation of Wright's within-species allelic imagery might be consistent with, but inappropriate as a simple depiction of, among-species evolutionary phenomena. Put another way, the nature (direction and intensity) of selection within populations and species produces a pattern of within-species temporal variation which may well be different from, or even opposed to, a net pattern of change resulting from differential species survival. The assumptions in Simpson's approach ignore this possibility entirely.

As already mentioned, Simpson (1944, especially p. 197 ff.; 1953) recognized three modes, or styles, of evolution. He pointed out that his classification simplified the complexity of evolutionary phenomena and he expressed the hope that it would serve as "one valid classification of basic descriptive evolutionary phenomena" (1944:197). Labeling the "practical study of changes in adaptation" as "the most essential single phenomenon of modes of evolution" (1944:189), Simpson explicitly relates his three modes to the "adaptive grid." Each mode, moreover, is assigned a "level" (in terms of the Linnaean hierarchy, and apparently phenomenologically) where it is most typical, though Simpson claims that each mode is not necessarily restricted to its most general level. Using the concept of "adaptive zones" ("consideration of the environment as composed of a finite, and a more or less clearly delimited set of zones or areas"; 1944:189), Simpson's tripartite classification of evolutionary modes follows [see figure 6.4, a reproduction of Simpson's (1944) figure 31, p. 198].

1. Speciation: differentiation, usually a within-species phenomenon. In terms of the adaptive grid, it involves either differentiation of a population into subzones of a single zone, or the elaboration of new adaptations, allowing later invasions into new subzones.

2. Phyletic evolution: the "sustained, directional (but not necessarily rectilinear) shift of the average characters of populations," a mode "typically related to middle taxonomic levels, usually genera, subfamilies, and families. In relation to the adaptive grid, phyletic evolution is usually or most clearly seen as a progression of single or multiple lines within the confines of one rather broad zone" (Simpson 1944:203).

3. Quantum evolution: "the relatively rapid shift of a biotic population in disequilibrium to an equilibrium distinctly unlike an ancestral condition" (Simpson 1944:206). Like other modes, it can give rise to taxa of any rank, but Simpson proposed quantum evolution as "the dominant and most essential process in the origin of taxonomic units of relatively high rank, such as families, orders, and classes." In terms of the adaptive grid, quantum evolution pertains to interzonal shifts.[3]

3. Crucial to our argument here is the demonstration of a direct relationship between Simpson's version of Wright's topographic landscape and his own concept of the "adaptive grid," which he characterized as follows: "The course of adaptive history may be pictorialized as a mobile series of ecological zones with time as one dimension. Within the limits of the flat pages, the basic picture resembles a grid, with its major bands made up of discrete smaller bands and these ultimately divided into a multitude of contiguous tracts" (Simpson 1944:191). That the two sets of imagery were related emerges particularly from Simpson's discussion of quantum evolution, as we shall discuss.

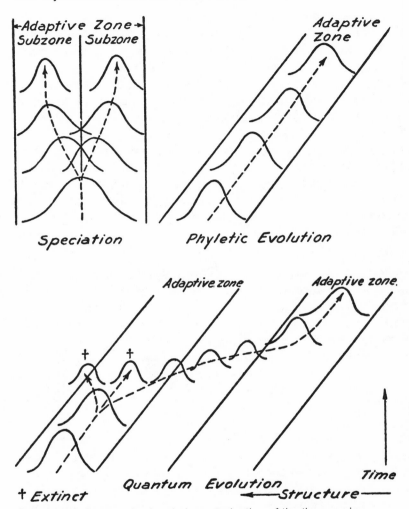

Figure 6.4 Simpson's visual characterization of the three major modes of evolution, in which "broken lines represent phylogeny and the frequency curves represent the populations in successive stages." (From Simpson 1944:198.)

The point of this brief recapitulation of Simpson's three modes is to show their total dependency on notions of adaptation, with a concomitant, underlying selection argument. Simpson apparently viewed speciation per se as an epiphenomenon; in any case, we have already presented our views on speciation and phyletic evolution, and shall further discuss the relative importance of these modes to the evolutionary process. The mode that Simpson proposed as really critical in macroevolution—quantum evolution—is a rather brilliant conception designed to explain just how the distance between Simpson's version of adaptive peaks is in fact traversed. As stated, it is purely a transformational postulate. The equation of the pictorial concept of the "adaptive peaks" and Simpson's own concept of the "adaptive grid" is neatly shown by Simpson's (1944) second discussion of the evolution of equid hypsodonty (see figure 6.5, a reproduction of Simpson 1944, figure 34). Simpson says (1944:209): "In discussing equid hypsodonty it was shown that there are two main adaptive zones, each embracing a shifting adaptive peak—one corresponding with browsing habits and one with grazing habits." There follows a long restatement, in terms of the quantum evolution mode, of the equid scenario quoted above. He then states (p. 210): "On a relatively small scale, the distance from the browsing to the grazing peak in this example is a quantum, a step that must be made completely or not at all, although in the example and, I think, in general the genetic processes involved do not permit making the step with a single leap,[4] and the selective processes do not make the unstable intermediate forms inviable."

Simpson later (1953, 1959) modified details, but did not alter his classification of the three modes. And thus has the foremost American student of macroevolution firmly equated the origin of taxa of higher categorical rank, in essence, with the problem of new and changing adaptations—a problem of getting from one peak to another.[5]

4. This remark represents an evident side-step from Goldschmidt's (1940) concept of "systemic mutations." Goldschmidt saw the problem of applying concepts of within-population variation and selection as a direct explanation of macroevolutionary phenomena (to the point of viewing speciation as a pure epiphenomenon, as did Willis 1940), but also saw macroevolution purely as a problem of morphological and genetic change—hence his ad hoc hypothesis of systemic mutations. Gould (1977a) and others have recently resurrected Goldschmidt's concepts (via regulatory genes, for the most part), an interesting and possibly fruitful action, provided the mechanism is located within the correct phenomenological level, obviously that of developing individuals within a population. It is not our purpose to deny that evolutionary theory must offer an explanation of how morphological and genetic change occurs, but rather to establish that such explanation is frequently not made within the proper context.

5. As is clear in the quote in the last paragraph (pertaining to the quantum evolution of equid hypsodonty), although Simpson repeatedly referred to it as especially concerned with "higher categorical levels," quantum evolution is a phenomenon that can only logically take place at the population and species level and can involve at most a series of species. Thus quan-

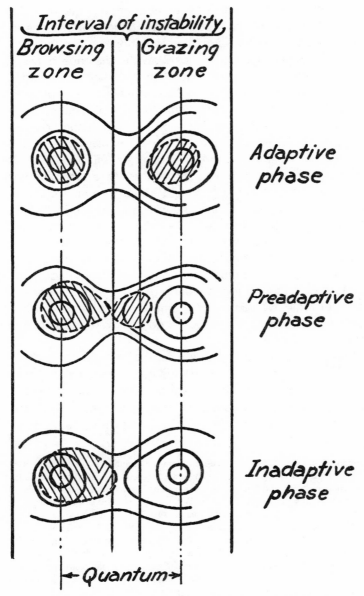

Figure 6.5 Simpson's "phases of equid history interpreted as quantum evolution. . . . The phase designations refer to the part of the population breaking away and occupying the grazing zone." (From Simpson 1944:208.)

Simpson (1959) returned to the topic of macroevolution, repeating his "conviction that the basic processes are the same at all levels of evolution, *from local populations to phyla, although the circumstances leading to higher levels are special* and the cumulative results of the basic processes are characteristically different at different levels" (Simpson 1959:255). The italics are ours, emphasizing again the confusion of phenomenological levels; the mechanisms Simpson cites can only be proposed to operate within populations. Quantum evolution is not mentioned in this later paper, but Huxley's (1958) concept of the "grade" is added to his considerations, and the discussion remains firmly rooted and wholly steeped in adaptation. There is a greater emphasis on splitting phenomena (producing adaptive radiations, presumably by speciation), but the fundamental conclusion,[6] that "higher categories generally arise by acquisition of a basic general adaptive complex" (Simpson 1959:270), continues the theme that evolution is a matter of adaptation via selection which goes on at the within-population level and on up through that of phyla.

Huxley (1958:27) wrote of grades as being "just as 'natural' or at least non-arbitrary as the customary monophyletic levels." They are levels of *anagenetic advance*—i.e., groups of organisms defined by some organizational (usually structural) improvement. Improvement, to Huxley, "covers detailed adaptation to a restricted niche, specialization for a given way of life, increased efficiency of a given structure or function, greater differentiation of functions, improvement of structural and physiological plan, and higher general organization.

tum evolution is, at base, a special aspect of speciation, an obvious consideration which may explain why Simpson (1976:5) and Boucot (1978) have claimed that "quantum evolution" is the early and exact equivalent of "punctuated equilibria" (Eldredge and Gould 1972); the latter authors equate punctuated equilibria with speciation. Earlier, Simpson (1953:389) stated that quantum evolution is "a special, more or less extreme and limiting case of phyletic evolution," i.e. not speciation. In any case, those portions of the discussions of "punctuated equilibria" pertaining explicitly to modes and rates of morphological change deliberately avoid use of terms such as "adaptation" or "natural selection" and are certainly not concerned with ways and means of getting from one adaptive peak to another.

6. We use the word "conclusion" loosely here. As Van Valen (1978) notes, Simpson's ideas have a way of retaining their plausibility and appeal despite the fact that the "data" upon which they are supposedly based have become outmoded or been superseded or outright falsified. Thus the genealogical hypotheses cited by Simpson as partial corroboration of his views are not, in fact, used to test the ideas at all. Rather, the ideas exist (perhaps originally suggested by patterns of data), and examples that appear to fit are selected to illustrate the concept. This contrasts with the approach we develop later in this chapter.

At all levels it is the direct consequence of natural selection" (Huxley 1958:19). This paper had a great influence on subsequent thought on macroevolution. Many authors were quick to point out that such and such a taxon was not, in fact, a natural monophyletic group (i.e., a clade), but was, rather, a grade—a non-monophyletic group defined on the basis of convergences or on non-homologous "pseudosynapomorphies." Simpson (1959) himself wrote a paper claiming that the Mammalia were patently polyphyletic and that the defining character (in this case, the anatomy of the jaw articulation) of Mammalia was not homologous in all mammals. The multiplicity of papers on grades in the late 1950s and early 1960s points up two things: (1) Many of the groups in conventionally accepted taxonomy of organisms were not monophyletic; elimination of not-A sets (chapter 5) begins with their recognition, so the concept of grades marked a healthy turn of events for systematics, and (2) The concept of adaptation via selection, a view central to most previous discussions of macroevolution, was so completely dominant that the ultimate extreme was reached: generally thoughtful biologists could seriously consider the supposed evolutionary processes leading to non-monophyletic (hence nonexistent in a genealogical, or evolutionary sense) groups.

The adaptational argument extends in detail to other aspects of macroevolutionary theory, involving directions of (morphological) evolutionary change as well as evolutionary rates. Perhaps the most classic, long-standing extrapolation of microevolutionary (within-species) phenomena to among-species (macroevolutionary) phenomena involves directionality. In no other aspect of evolutionary theory is the confusion of within- and among-species phenomena more clearly seen. The literature on directional phenomena in evolution is enormous, but there is little to be gained from a detailed consideration of particular examples. Suffice it to say that there is a fundamental assumption, pervasive in neo-Darwinian theory, if not held by all neo-Darwinists, that the same phenomenon—fitness differences among individuals within a population, or natural selection, which by definition alters the complexion of the gene pool from generation to generation—can, when viewed over evolutionary time, be seen as a cumulative process with rather large net effect. Among paleontologists, Gingerich (e.g., 1976, 1979) has been most eloquent in recent years in defense of this view. And nowhere can a more careful, thorough presentation of the transformation of entire populations, forming single lineages or "phyla," be found than in Simpson's (1944, 1953) discussions of phyletic evolution. It is to be noted that Simpson's belief—at least in 1944—is that phyletic evolution is most evident at the "middle levels" of genus, subfamily and family, a sort of "low macro." Directionality, including, but not restricted to, linearity, is generally considered the direct consequence of directional selection (e.g., Simpson 1953; Hecht 1965). Thus trends,[7] especially in morphological features of fossils collected from successive horizons, are most commonly considered as an example in geologic (or true evolutionary) time of exactly the same phenomenon on the generic and familial level as seen in directional selection experiments within *Drosophila* population cages. Eldredge and Gould (1972, 1974) and Gould and Eldredge (1977) have characterized this particular set of ideas at length. In addition to the theoretical reasons why directional selection would not as a rule be expected to persist over millions of years, it is apparent (see p. 283 ff.) that data pertaining to within-species variation in most examples of trends in the fossil record in effect falsify the hypothesis that among-species phyletic phenomena, including trends, are a direct reflection (and hence outgrowth) of within-species trends that might plausibly be attributed to natural selection. (The reader is referred to these papers, and references cited therein, for further discussion of these points.) An alternative view, that among-species (and, of course, taxa of higher rank) trend phenomena result from a process of differential species survival and are not directly attributable to within-species patterns of temporal variation (whether directional or not) has been available in the literature for many years, but until

7. All patterns, including trends, discussed in this chapter are considered to have at least three components: (a) a deterministic component. We are interested specifically in the deterministic element of trends at this juncture; (b) a random component. We discuss stochastic processes and their role in the production of patterns, especially on page 298; (c) an error component. There are many sources of error in the perception of pattern. In this connection, we note earlier claims (e.g., Eldredge and Gould 1972:111) to the effect that "many, if not most, trends involving higher taxa may simply reflect a selective rendering of elements in the fossil record" (see also p. 323). Salthe (1975:302) has gone further, suggesting that trends ("series") have no ontological status whatever, being in all cases solely creations of evolutionary biologists. We here assume that at least some major component of the patterns we call "trends" are deterministic in origin, i.e., not only that the patterns reflect an actual shift in value of some specified intrinsic feature during the genealogical history of a group, but also that at least some such shifts are the result of deterministic processes.

recently has been by far subordinate to the conventional view of phyletic modification of entire populations via the direct action of natural selection.[8]

The transformational view of macroevolution has also dominated discussion of evolutionary rates. Under the conventional definition of evolution ("changes in gene content and frequency") it is natural that evolutionary rates have been taken to mean "rates of change of gene content and frequency." Taxonomic rates, a topic first fully developed by Simpson (see 1944 and, especially, 1953) have frequently been used as a means of estimating rates of genetic change. Rates of diversification of taxa of higher categorical rank are common in the literature of paleontology; frequency curves showing the number of new taxa first appearing, or disappearing, or taxa persisting (the "standing crop") during an increment of geologic time are handy devices to summarize fluctuations in diversity of a group through time from the point of view of a given taxonomic level. It is unclear what else such rates of origin of higher taxa might mean, except as successively poorer approximations to actual speciation rates, the higher up the Linnaean hierarchy one goes.

If taxonomic rates are viewed as a means of estimating actual genetic rates, it is perhaps, but not necessarily, true that there is a hidden assumption of adaptation/selection working equally at all taxonomic levels, i.e., that taxonomic rates are a function, ultimately of

8. In his discussion of *Stufenreihe,* or "steplike" evolution, Simpson (1944:194, figure 29; 1953:220) notes that many, and perhaps most, apparent trends reflect "relatively abundant, relatively static populations of successively occupied adaptive zones" (1944:195). As shown in his figure 29, the change within each "structural stage" is irrelevant to the overall direction of the trend. The phenomenon is not discussed at length, and structural stages (i.e., grades) are considered, rather than clades, but there nonetheless is a deviation from wholesale extrapolation of within-population and within-species patterns of (temporal) variation as a direct correlate of, and underlying mechanism for, directional among-taxon patterns (trends). Van Valen (1978:211) claims, without citation, that the notion of lineage selection has a long history, being used explicitly by Lyell, Darwin, Simpson and himself. To which it may be remarked that, in the works of authors as prolific as these four, there is usually to be found a little bit of virtually anything at all relevant to the topic. This is as it should be. But certainly, the main gist of Simpson's writings, to take one example, is far removed from any general notion of "lineage selection." Van Valen (1978:211) further states that such notions somehow imply that evolution *within species* is unimportant, whereas it is just the other way around: the only phenomenological level at which adaptation (via selection) actually can occur is within species. It is one thing to postulate that once a species has developed its own set of adaptations, selection tends to conserve it and quite another to state that within-species evolutionary phenomena are not important!

morphological rates, which are in turn a function of genetic rates. As will be developed below, the converse possibility—that genetic and morphologic rates might, in some sense at least, reasonably be hypothesized to be a function of speciation rates—emerges from the recognition of the dichotomy between within- and among-species evolutionary phenomena.

But the main point concerning rates is that, at base, it can be entirely appropriate to compare rates of genetic change among monophyletic groups of any rank, regardless of magnitude. This is because hypotheses of evolutionary mechanisms are fundamentally a matter of mode, i.e., styles of change, including recognition of units of change and biotic and physical parameters moderating such change. To assert that species evolve, not anatomical parts, or allelic loci, of individuals, is not to deny that changes in gene content and frequency result from the evolutionary process. As long as considerations of rate imply no necessary mode of change (e.g., phyletic transformation vs. speciation), relative rates of genetic change among phyla may be studied. The problem of mixed phenomenological levels arises only when comparative studies of morphological and genetic rates are used to deduce generalizations about the actual nature of the evolutionary process.

Transformational Macroevolutionary Theory: Summary and Critique

The concept of natural selection (fitness differences, or differential reproduction of individuals within populations) appears to be a corroborated, within-population phenomenon, and constitutes the best available explanation for the origin, maintenance, and possible modification of adaptations. Its corroboration results primarily from laboratory and (mathematical) theoretical studies of populations. In bridging the gap from within-species phenomena to patterns of taxonomic and morphological diversity of the earth's past and present biota, these principles have been applied wholesale across the taxonomic hierarchy, as though there were no difference between a population, on the one hand, and a kingdom, on the other. There are two fundamental objections to this approach, both arising from a failure to integrate notions of the nature of species and their modes of origin fully into the paradigm. First, if species are viewed as individuals (i.e., discrete entities with origins, subsequent histories, and definite terminations), their evolution must be explained. Only a view that species are transitory, arbitrarily defined segments of an evolutionary continuum permits the notion that within-population phenomena may be extrapolated directly to higher levels. Recognition of the existence of species as discrete entities in effect contradicts the vision of change in gene content and frequency—whether or not effected by natural selection—as a continuous process from the population on up through the phylum.

The problem is particularly acute in terms of the concept of adaptation. We accept, for purposes of discussion, the basic notion of adaptation, the adjustment of intrinsic features in response to natural selection. As developed, it is a within-population, generational phenomenon. The problem arises from speciation theory: *there is no necessary relationship between natural selection and speciation,* at least in terms of allopatric speciation (see Eldredge 1974a:542; "Though selection invariably plays an important role during any speciation event, it has never been shown to be the effective "cause" of speciation in the sense that the selective regime originates and is maintained for the "purpose" of developing a new species"). As reviewed in chapter 4, Bush has pointed out that selection for reproductive isolation is a necessary component of sympatric and parapatric speciation. But nowhere in contemporary works on speciation theory is the notion developed that speciation is fundamentally a process of adaptation. New adaptations, or the perfection of old ones, might be acquired, particularly in situations involving relatively small, peripherally isolated populations. But such adaptations, especially in allopatric situations, are incidental to the major phenomenon of the establishment of reproductive isolation. Any change involving behavior, morphology, or cytogenetics may be sufficient to effect reproductive isolation. Such change may be adaptive as well, but to conclude that speciation is a phenomenon of adaptation is a distortion of contemporary speciation theory. Speciation is a matter of establishment of reproductive isolation; adaptation via selection may or may not be involved incidentally. And, at least in the allopatric case, it cannot be construed that selection itself acted to create two species from a single ancestor, i.e., that the prime "object" of selection was the actual creation of two species from a single ancestor.

From this point of view, the direct extrapolation of the concept of adaptation via selection to explain differences in intrinsic features among taxa of higher categorical rank cannot be appropriate. The use of the adaptive landscape beyond the limits of a single species, with its fundamental underlying premise that the within-species phenomena of selection and adaptation can be extrapolated in direct, unbroken fashion, violates the notion that the origin of new reproductive communities (speciation) is not logically to be considered a phenomenon of adaptation. Rather, speciation breaks the smooth, within-population generational process of adaptation via selection. Thus the entire landscape metaphor is inappropriate for purposes other than the one for which Wright originally conceived it: within-species patterns of distribution of relatively more "harmonious" combinations of alleles.

The second objection is that taxa of rank higher than species do not exist in the same sense as species exist. Species are reproductive communities. Taxa—at least properly defined, monophyletic taxa—consist of one or more species connected by unity of descent. That is the definition of the word "taxon," holding for orders, classes, and taxa of all other categories. This consideration suggests that genera, orders, and so forth, do not evolve except as their component species do—that the patterns of fluctuation of diversity within taxa of higher categorical rank are a reflection of patterns of origin, survival, and extinction of their constituent species.

Thus the notion of species as discrete units simultaneously disrupts the ebb and flow of population continua (based on continuous, generational change) and collapses the study of macroevolution down to a single phenomenological level. When nature is viewed as a dynamic, functional system, the next step above the level of species is the ecosystem, not the genus, to paraphrase Salthe (1975 and in press), who uses the terms "population" and "community" for "species" and "ecosystem," respectively. (See Valentine 1969, for a similar view of natural hierarchies.)

This brief characterization of contemporary transformational macroevolutionary theory has been eclectic and was written primarily to document the evident confusion in phenomenological levels that has plagued the subject. To document the wholesale extrapolation of microevolutionary phenomena across to higher (Linnaean) levels, we have chosen Wright's concept of the adaptive landscape and some of the uses to which it has been put. We chose the landscape imagery partly because it has played a crucial role in both the analysis and the pictorial representation of these diverse phenomena in terms of adaptation and selection, and partly because Wright himself avoided the pitfall of such bald extrapolation. The landscape metaphor continues to provide the point of departure for most discussions of transspecific phenomena right up to the present day (see, for instance, the chapter on transspecific evolution in Dobzhansky et al. [1977] and the somewhat equivocal usage of the imagery by Lewontin 1978). We must agree with Waddington (1967:14) who said: "The whole real guts of evolution—which is, how do you come to have horses, and tigers, and things—is outside the mathematical theory [i.e., within-populational phenomena of change in gene content and frequency]." This is not because that theory is wrong, but because there are indeed discrete phenomenological levels glossed over by such across-the-board applications of population-genetics theory to among-species phenomena.

We have also parenthetically noted that most of the deviation from orthodox syntheticism has been equally plagued by the notion that the essential problem of macroevolution is the explanation of morphological differences, and that the majority of such efforts have also ignored the central importance that species have in the problem. Other biologists, as we shall soon see, have perceived the same set of problems. Moreover, resolution of the problem involves clarification of the phenomenological levels themselves. The actual ingredients of a revised theory of macroevolution consist of long- and well-understood biological principles, and require no invention of new mechanisms. Our next task, after reviewing earlier work, is to set forth an alternative approach to the problem.

The Phenomenological Levels of Evolution and their Relation to Systematics

Previous Work

Dissatisfaction with the concept of natural selection as the process of evolution has been evident since its initial espousal by Darwin and Wallace in 1857. Most criticisms have thrown the baby out with the bath water; of greater interest are statements and hypotheses which explicitly or implicitly recognize the phenomenological levels of evolution, particularly those which have promulgated a concept of intergroup selection.

Darwin's (1859) subtitle to his *Origin of Species,* "Or the Preservation of Favoured Races in the Struggle for Life," implies a concept of group selection (differential group survival) which was not, in fact, strongly developed within the book itself. Nonetheless, over the intervening years, hypotheses of intergroup selection have continued to crop up (rather like hypotheses of sympatric speciation) and have periodically been attacked by biologists asserting the purity and primacy of natural selection *sensu stricto*—differential reproduction among individuals. Within the last 50 years or so, the time period with which we have been most concerned in terms of the historical development of evolutionary theory, concepts of group selection have pertained mostly to interdemal phenomena. Interspecific competition and differential persistence—"species selection"—have fared poorly as general concepts, warranting, for example, a single sentence in an otherwise generally excellent recent text on evolution.[9]

Largely ignored have been the occasional remarks of Wright over the years. For instance, Wright (1931) clearly recognized that species are the units of evolution ("the evolutionary process is concerned, not with individuals, but with the species" 1931:98) and wrote briefly (p. 154) of intergroup selection (interdemal) as being important "in the origin of peculiar adaptations and the attainment of extreme perfection." In the paper in which he elaborated the notion of the adaptive landscape (Wright 1932) containing the sentence quoted earlier in this chapter (p. 252), intergroup selection is repeatedly brought up, but again with emphasis on interdemal selection within species. And in his review of Simpson's (1944) *Tempo and Mode in Evolution,* Wright (1945:416) expressed support for the view that intergroup selection is "creative," illustrating his point with reference to the elimination and "compensatory adaptive radiation" of "families and orders of vertebrates," clearly expanding his view of group selection to the species level and far beyond—though he

9. "Related species compete for resources that both are in need of, and one species may outbreed and crowd out another" (Dobzhansky et al., 1977:125).

stressed that his main point remained inter-demic selection. He then went on (1945:416 ff.) to discuss interdemic selection based on altruism, the arena wherein much contemporary discussion (starting with Haldane 1932, according to Williams 1966:92) of group selection persists. Wright (1956:21 ff.) further discussed selection "among non-interbreeding species and higher categories," pointing out that (p. 22): "The raw material for such selection is even less random in an absolute sense than in the case of interdeme selection since it consists in the differentiation of these categories that has come about as a result of all of the preceding evolutionary processes. The competing groups may, however, be looked upon as random trials from the standpoint of the course of evolution of life as a whole that comes out of their competition." He further writes: "if there were no selection between such categories, we would expect to find all of those of an early geologic period persisting and all branching to similar extents." In other words, the very nature of the historical record indicates that differential species survival is a self-evident fact. He concludes his paper (Wright 1956:23) with the following remark: "The course of evolution of vertebrate life and of life in general has been guided throughout by a hierarchy of processes of selection ranging from selection between genes to selection between orders, classes, and even phyla." We would, of course, demur only with the claim that differential survival of clades of rank higher than species represents anything more than differential species survival.

Finally, as already quoted and extensively discussed by Eldredge and Gould (1972:111–12), Wright (1967:120) wrote: "With respect to the long term aspects of the evolution of higher categories, the stochastic process is speciation. This was treated as *directed* above but may be essentially *random* with respect to the subsequent course of macro-evolution. The directing process here is selection between competing species often belonging to different higher categories." Eldredge and Gould (1972:112, figure 5–10) utilized this notion as the basis of their reconciliation of the existence of long-term trends (i.e., net, directional changes in one or more morphological features within a monophyletic group through geologic time), with their rejection of the hypothesis that net, within-species variation is usually, or even commonly, directional through time. Stanley (1975) clarified and expanded this line of argument, referring to "species selection," and concluded that inter-specific evolutionary phenomena are in fact "decoupled," from intra-specific phenomena. Gould (1977b) and Gould and Eldredge (1977) used the expression "Wright's Rule" to refer to this aspect of the decoupling of phenomenological levels.

Similarly, Grant (1963:397; 1977) has used the term "interspecific selection" as an alternative to the more conventional hypothesis of "orthoselection" as an explanation of directional, interspecific trends within monophyletic groups. Mayr (1963:586), addressing the topic of "transspecific evolution," began his chapter by writing: "The proponents of the synthetic theory maintain that all evolution is due to the accumulation of small genetic changes, guided by natural selection, and that transspecific evolution (Rensch 1947) is nothing but an extrapolation and magnification of the events that take place *within* populations and species" (italics added). However, in the last three paragraphs (p. 621) of his chapter on transspecific evolution, Mayr clearly states the notion of speciation supplying the raw material of variation for macroevolution: "Species, in the sense of evolution, are quite comparable to mutations." This view is identical to Wright's (see especially 1967) and at variance, for reasons we have already developed at length, with the statement (quoted above) with which Mayr opened his chapter.

Hull (1976) has discussed hierarchical levels in conjunction with the notion that species are individuals but leaves open the question (p. 184) whether monophyletic higher taxa "possess sufficient unity" to function as "units of evolution." His conclusion on the topic follows: "If macroevolutionary change is more than a summation of microevolutionary events, then complexes more inclusive than species might also form units of evolution and count as individuals." Although we disagree that a view of macroevolution as "more than a summation of microevolutionary events" leads to the conclusion that taxa of rank higher than species are themselves individuals and "evolutionary units," Hull's arguments on the status of species as individuals have advanced the general concept of different levels of evolutionary phenomena.

Apart from such theoretical statements and hypotheses, there has been relatively little explicit analysis of macroevolutionary problems in which within- and among-species phenomena are clearly differentiated. One example of the use and power of such discrimination can be seen in Bock's (1970, 1972) analysis of the "adaptive radiation" of the Hawaiian honeycreepers (Aves, Drepanididae). In contrast to his earlier approach (Bock 1965)[10] to the origin of birds, in which the emphasis was actually on the adaptation of avian flight and a gradual model of successive stages was developed, Bock's analysis of honeycreeper evolution (see figure 6.6) involved: (a) a theory of relationships among known (i.e., extant) species of Hawaiian honeycreepers. Although no cladogram was presented, there was clearly developed a hypothesis of degrees of relatedness among the species, coupled with a hypothesis of primitive and derived morphologies, and generalized vs. specialized feeding types. The analysis proceeded despite the acknowledged possible absence of extinct (unknown and unknowable) species from the sample. The analysis further involved (b) explicit invocation (especially in Bock 1972) of the concept of interspecific competition as the source of the "selection force" underlying the origin of new morphologies. The study is an excellent example, at least at the hypothetical level, of the recognition of separate phenomenological levels and discussion of the production of the hypothesized pattern of evolutionary relationships among species of Drepanididae explicitly in terms of the component species sampled from that family. Testability of such hypotheses is considered in more detail below.

There are, undoubtedly, still further examples wherein the notion of distinct levels, with species and speciation forming the crux of the matter, has been developed. However, it is manifestly clear that the theme is not a dominant one in contemporary evolutionary theory. We believe it should be.

10. As this book was going to press, Bock published a review of macroevolutionary theory (Bock, 1979). His position in this paper is thoroughly and exclusively transformational. Although he initially defines macroevolution (p. 20) as "the appearance and subsequent specialization of distinctive new features and taxa," neatly combining and confounding the transformational and taxic elements, the remainder of the paper views macroevolution as "simply a large amount of phyletic evolutionary change" (Bock 1979:36, under the heading "Speciation and Macroevolution"). To accomplish his avowedly reductionist task of explaining macroevolution strictly within the terms of microevolutionary processes, Bock (p. 28 ff.) concludes that the biological species concept [e.g., one like Mayr's (1941), cited here on p. 92] has no time dimension: phyletic lineages exist, but have no properties as discrete units or entities in time. Denying the ontological status of species as real, discrete units in time and space is, of course, a logical necessity for all neo-Darwinian or syntheticist versions of transformationalism (see Eldredge 1979b:8, for a discussion of this point). Bock's conclusion is therefore inevitable (1979:39): "The key to all explanatory models of macroevolution is the concept of biological adaptation." We conclude, in contrast, that Bock's paper (1979), although citing his work on the Depanididae, fails to appreciate its full significance and instead reverts to the more classically conceived transformational approach seen in his earlier work (e.g., Bock:1965).

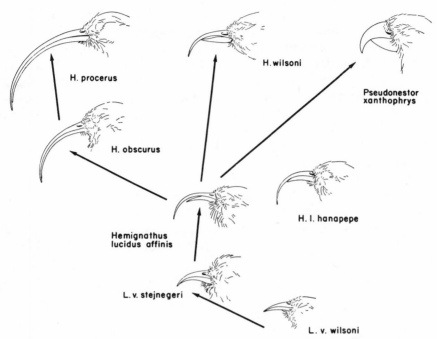

Figure 6.6 Diagram illustrating head morphology in profile and hypothesized phylogenetic relationships of species within three genera of Drepanididae. (From Bock 1970:714.)

Methodological Considerations

We have emphasized our view that the very existence of species directly implies a hierarchical organization of nature. The properties of species, as discrete entities, effectively "decouples" (Stanley 1975) within-species from among-species evolutionary phenomena. Thus there are two phenomenological levels to evolution: within-species change in gene content and frequency ("microevolution") and change in species composition within a monophyletic group in space and time ("macroevolution").

Origin of new species from old (speciation) is the mechanism which effectively decouples the two levels of microevolution and macroevolution. It should further be noted that other phenomenological levels are also pertinent to evolutionary theory. At the level of the individual, for example, it is obvious that the modification of developmental pathways underlies all change of intrinsic features in phylogeny. At the cellular level, the physicochemical constraints and inducers of codon changes become relevant. And so on. Any complete theory of evolution must deal with all definable levels, as well as with their independence and interdependence. In this book, we confine our attention to micro- and macroevolution, for the simple reason that the data of systematics are pertinent to these levels. Indeed, as we shall soon see, in terms of the analysis of processes acting within levels, the data of systematics pertain solely to differential species origins and extinctions—the dynamics of macroevolution.

Microevolution We have already commented at length that the twin neo-Darwinian concepts of adaptation and selection, which arise from and are appropriate to the population level (i.e., within-species level) have been inappropriately extrapolated to the higher level of macroevolution. We have also expressed our agreement that natural selection, despite persistent criticism, is an actual dynamic process in nature. Within populations there is inevitably a sampling of the available genetic variability of a given generation represented in the succeeding generation. Some of this sampling, at least, may be non-random—a reflection of differential reproductive success within the parental generation, linked in some way to relative survival value of a given behavioral, physiological or morphological trait.

But natural selection is strictly a within-population phenomenon by definition (see Salthe 1975). Nearly all criticisms of natural selection are based, in the final analysis, on the perception that explanations of evolutionary phenomena frequently invoke natural selection at the wrong phenomenological level. Natural selection should only be hypothesized under conditions in which it can be tested directly. Data required are comparative gene frequencies in a parental and one or more descendant populations. By definition, the concept is designed for, and limited to, within-population situations. It should not be surprising, therefore, that the concept can only be applied by the evolutionist in such situations. At higher levels, the use of a concept such as natural selection (e.g., as in Simpson's equid scenario cited and quoted above) is inappropriate, both conceptually as just discussed, and also epistemologically: in an investigative protocol which minimally demands that hypotheses be susceptible to criticism, the concept of natural selection—if applied to any situation (i.e., level) other than that for which it is appropriate—fails utterly. Applications of the concept of natural selection to among-species phenomena are therefore inappropriate on methodological grounds because the data are themselves not appropriate to evaluate the specific hypothesis at hand.

Thus it follows that there are also methodological grounds for rejecting *adaptation* as a fruitful way of conceiving of and addressing issues in macroevolutionary theory. This statement holds, of course, only if adaptation is strictly and exclusively tied to natural selection, a conventional view which we find highly corroborated in past considerations of the subject. Accepting this relationship, it would appear that adaptation—adjustment of behavioral and anatomical traits to perform explicit functions with respect to biotic and abiotic parameters of the environment—is effected at the within-species level. Inasmuch as it is populations which are integrated into ecological communities, the conclusion seems ineluctable that study of the process of adaptation (true, evolutionary adaptation—not just physiological adaptation which goes on at the level of the individual) is fundamentally an ecological and experimental problem. For its scientific study, genetic data and well-corroborated hypotheses of functional morphology on successive generations over a number of years are required. The requirements are, therefore, stringent, but perhaps not impossibly so. The process of adaptation simply cannot be applied to, or studied from the vantage point of, the higher phenomenological level of the evolutionary process that we call "macroevolution." For these reasons, in addition to those cited above, we should therefore drop adaptation as the focus of our research into macroevolution. Adaptation (via natural selection) is a most important process in evolution—but it is a within-population generational phenomenon that requires data unavailable to all paleontologists and to most systematists as well. As Salthe (unpublished manuscript) has remarked, the dynamics of within-population phenomena are the best understood of any of the hierarchical levels; they are best left to population biologists, especially geneticists and ecologists, and will not be discussed further here.

Speciation The origin of the reproductive units we have been calling species has already been discussed from the standpoint of systematics earlier in this book. We developed our views on the nature of speciation at that juncture because of the evident link between notions of speciation, on the one hand, and the construction of phylogenetic trees on the other. To convert a cladogram into a tree, one must specify ancestors at branching points (chapter 4) and a discussion of modes of speciation is of heuristic value, relevant to an analysis of the numbers and nature of phylogenetic trees that may be obtained from any given cladogram.

Here, we briefly stress what is, in effect, the converse: that the study of speciation requires, among other things in its data base, a highly corroborated phylogenetic tree linking all taxa (species, "subspecies," and perhaps even populations) in some definite ancestor-descendant pattern. In other words, a detailed reconstruction of the historical relationships of two or more taxa at or below the specific level is the *sine qua non* of all analyses of speciation. Temporal and geographic distributional data—or hypotheses—are also requisites.

Although most studies of speciation list many examples (patterns of relationship and distribution which appear to confirm or illustrate particular aspects of speciation theory under discussion), in point of fact, the only means whereby speciation theory can be tested adequately is by comparison of predictions arising from theory with patterns (trees plus distributional information) worked out independently from that theory. That this is a tall order quite difficult to fulfill in most instances arises both from the intrinsic difficulty of testing phylogenetic trees themselves (chapter 4) and from the near-impossibility of constructing these trees without reference (conscious or unconscious) to some theory of speciation or another. As we have already discussed extensively in chapters 3 and 4, one's very concept of species limits the choice of speciation theories to one subset or another of all those available and makes it almost certain that such a theory will be held a priori in the mind of one who is trying, nonetheless, to study speciation.

Thus, study of speciation offers some interesting contrasts to the levels of microevolution below it and macroevolution above it. Superficially, the testing of speciation hypotheses requires "data" simpler in detail than those required for microevolutionary study: for the former, we need a highly corroborated tree plus distributional data, whereas for the latter, we need detailed, generation by generation data pertaining to changes in gene content and frequency. But given the highly dubious nature of phylogenetic trees, as hypotheses difficult to test (i.e., of relatively low corroborability), the data base for testing hypotheses of speciation is in reality a great deal more complex than that required for testing microevolutionary hypotheses.

Another contrast between the phenomenological levels would appear to be the order of magnitude of time required for the processes to work. At first glance, the most simple generalization would be that microevolution is a generational phenomenon, and speciation involves origins of new units from old, each made up, minimally of one, but realistically of hundreds and usually thousands of generations. Macroevolution, invoking differential survival of species within monophyletic groups as its dynamic process, involves all possible lengths of time, bounded only by the age of appearance of the first species of that clade. But intergenerational within-species (microevolutionary) phenomena, technically at least, go on and can accumulate for the entire length of time an individual species persists—many millions of years not, apparently, being uncommon, although we have noted (chapter 4) that, to judge from patterns of phenotypic modification, relatively little change tends to accrue in such instances. In contrast, speciation may take only a few years, and perhaps normally requires only a few thousand years, to take place. Thus there is no simple relationship between the phenomenological level and amount of evolutionary time required for events to occur, or for the additive effects to accumulate. Accumulation of macroevolutionary change takes longer than accumulation of microevolutionary change simply because the cumulative existence of two or more species is almost automatically greater than the duration of any single species.

We conclude that the scientific study of the process of speciation demands, as its data base, the existence of well-corroborated theories of relationship among populations, or closely related species (or allopatric taxa whose precise status is moot, e.g., "semispecies"). Problems in testing the trees themselves add an unavoidable element of uncertainty to all such studies, a methodological limitation on our ability to assess competing hypotheses. For this reason, speciation probably will never be as well understood as are the processes at the lower, within-species level of microevolution.

Macroevolution Among-species phenomena constitute the next higher phenomenological level above the within-species level. The dynamic process acting at this level is taken to be differential species survival within monophyletic taxa. Data required to test specific hypotheses of macroevolutionary phenomena, under this conception of the level, would appear to be temporal and spatial distributional data of component species, plus well-corroborated cladograms showing these species as members of a monophyletic group. For most hypotheses, detailed trees appear not to be required.

Inasmuch as there is only a relatively small number of published hypotheses of this nature available, it seems relevant at this point to consider two possible courses: (a) generalization ("laws") arising directly from patterns of differential species survival, or (b) invention of hypotheses of process which can generate predictions about pattern, against which such independently analyzed patterns can be compared. The first would be strictly an inductive process, while the second contains a deductive element. Whether the difference between the two is of any ultimate significance, the latter course, offering direct means of criticizing hypotheses, seems preferable, albeit more difficult.

That the testing of macroevolutionary hypotheses requires only a cladogram, whereas testing of those pertaining to speciation per se requires trees, may at first seem surprising, inasmuch as speciation would therefore require the seemingly more complex hypothesis of pattern. But taxa of higher rank than species do not exist in the same way as do species (p. 249), and, as a corollary, such "higher" taxa cannot logically be ancestral to one another. Thus, even on this level of analysis, it is a false issue to speak of the "evolution" of the Equinae from the Anchitheriinae. Thus we require branching diagrams involving species themselves. Since trees can only give us details of precisely how any two species may be hypothesized to be related, it would be superfluous, at least in most instances, to consider whether a pattern of differential species survival involves ancestors vs. descendants, rather than, say, "plesiomorphous" vs. "apomorphous" species. A further methodological consideration is that to formulate macroevolutionary hypotheses requiring trees would be, in effect, to require the assumption that all species that ever existed within the monophyletic group were represented in the available sample under study. We conclude that there are very real constraints on macroevolutionary hypotheses: to be testable at all, insofar as phylogenetic patterns are concerned, they must deal only with branching diagrams of corroborated hypotheses of monophyletic descent (cladograms) and they must deal with species.

There are some hypotheses at the macroevolutionary level which can be most conveniently tested with data pertaining to taxa of rank higher than species. Such data (e.g., those pertaining to appearance of new, disappearance of old, and "standing crop" of genera, families, orders, etc., usually given in some graphical form) are only directed to such hypotheses as here conceived when it is assumed that such data are successive approximations away from (the higher up the hierarchy one goes) actual species diversity data. Much has been written in recent years about the use of generic and familial data, which are thought to allow us to estimate true species diversity in the fossil record. Use of generic and familial data, it is argued, helps avoid the loss of direct sampling of many rare species with restricted temporal and geographic ranges, but is of low enough rank to minimize the distortion of true species diversity (see figure 6.7). (This consideration, of course, applies only to comparative diversity among groups with an equal and reasonable chance of fossilization in the first place.) When such data are presented as an estimate of species diversity data, they are legitimate to use in the evaluation of macroevolutionary hypotheses. As we have already noted, use of such data to evaluate inappropriate hypotheses regarding putative evolutionary processes among "higher-level taxa" is not legitimate.

Relationships among the Phenomenological Levels of Evolution: Decoupling

In this section, we shall reiterate the conclusion that the phenomenological levels of the evolutionary process are "decoupled," at least to the extent that direct extrapolation or inference from one level to another is not possible. We are aware that, by definition, natural selection is a within-population phenomenon, and thus cannot be applied appropriately to higher levels. At issue at the moment, however, is the counter position, the widely accepted reductionist thesis that the levels do not exist as such and that patterns of ma-

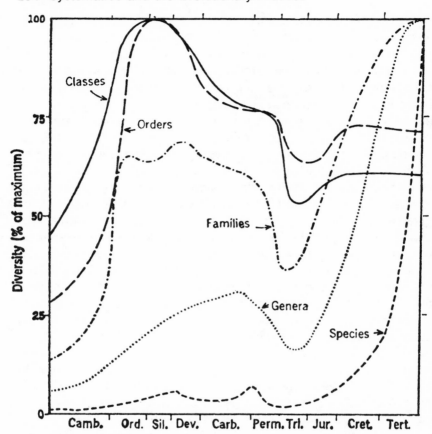

Figure 6.7 Comparison of apparent Phanerozoic diversity curves as measured at different taxonomic levels. Only well-skeletonized marine invertebrates were tabulated. (From Raup 1972:1070. Copyright 1972 by the American Association for the Advancement of Science.)

croevolution are a direct reflection of within-population dynamics of change.

Stanley (1975) has discussed at length the testing of these rival hypotheses ("decoupling" of discrete phenomenological levels vs. a direct continuity between within- and among-species patterns of selection). The best example of the latter form of hypothesis, phyletic gradualism, asserts, *inter alia,* that long-term (including interspecific, intergeneric, etc.) trends are the direct product of selection within species (true natural selection). Inasmuch as there are no data sets known to us that can definitely be shown to reflect selection within and across taxa (a possible example will be cited below, however), the hypothesis that the levels exist and are in fact decoupled might be restated as follows: patterns of within-species variation are not the same as those among species within a monophyletic group. This concept is illustrated in figure 6.8. If time-averaged patterns of variation within the span of existence of an ancestral species are of the same nature (have the same direction through time) as those of its descendant (figure 6.8a), the observation is consistent with the hypothesis that among-species differences arise smoothly and continuously from patterns of within-species variation. We hasten to add that identity or strong similarity of such directions does not utterly reject the existence of discrete levels. As Gould and Eldredge (1977) have argued, coincidence of within- and among-species patterns of variation within a hypothesized series of ances-

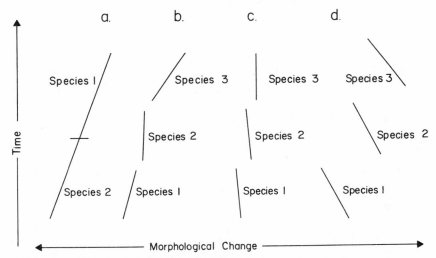

Figure 6.8 Relation between within- and among-species time-averaged patterns of variation. (a) A hypothetical situation in which within-species direction of change through time is continuous and unbroken across species "boundaries." This type of change is consistent with the hypothesis that within-species variation and direction of change are responsible for among-species patterns. (b) A situation in which the fossil record of three species shows within-species patterns similar to the total among-species trend, likewise consistent with the above-stated hypothesis, but also consistent with the hypothesis that the levels are decoupled. (c,d) Directed, net change (a trend) among ancestors and descendants where the net change among species is inconsistent with the hypothesis that such trends are the direct product of within-species patterns of directional change.

tors and descendants may still reflect a "decoupling" phenomenon at speciation, the resumption of the trend within the descendant species reflecting pure chance (the stochastic element, discussed particularly on p. 298 ff.), or the resumption of a pattern actually disrupted by the wholly unrelated event of speciation (figure 6.8b). If patterns of time-aggregated variation, especially in those characters used to distinguish the related species, are "neutral" (i.e., there is no net change in mean value, figure 6.8c) or in the opposite direction (figure 6.8d) of the pattern of net directed change among species, then the hypothesis that among-species differences arise as a mere, and direct, extrapolation and accumulation of within-species patterns is effectively rejected.

Such tests are inherently weakened by the basic assumption that directed patterns of net aggregated variation through time in fact reflect selection. They are further weakened because they deal with the critical level of speciation, and therefore ideally require highly corroborated trees. The proposition can perhaps be generalized to state that relatively apomorphic species within a monophyletic group should not exhibit patterns of within-species variation coincident with such patterns within plesiomorphic sister-taxa. If the hypothesis that within- and among-species patterns of variation through time are different is valid, then the pattern should hold for monophyletic groups whose relationships are depicted on cladograms. We now present three examples from the fossil record. In each case, taxa which are putative descendants, or simply putative apomorphic sister-species, show a pattern of within-species variation radically different from that of the putative ancestor or plesiomorphic sister-species.

Makurath and Anderson (1973) analyzed the phylogenetic history of the Upper Silurian and Lower Devonian species of the pentamerid brachiopod genus *Gypidula* in the Appalachian region. Basing their conclusions on statistical analyses of two morphometric variables (beak length and spondylium width), the authors concluded that over three successive stratigraphic horizons, evolutionary change of the two anatomical features was smoothly continuous both within and among the two species. Eldredge (1974b) criticized the experimental design of this study, concluding that the data were inadequate for a true test of the rival hypotheses of "gradualism" vs. "punctuated equilibria." Of interest here is the

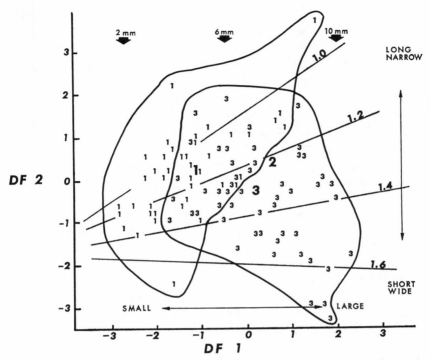

Figure 6.9 Within-sample patterns of variation of *Gypidula prognostica* (group 1) and *G. coeymanense* (groups 2 and 3). Specimen scores are plotted against the first and second discriminant functions. Large numerals are respective group centroids; small numerals are individuals. Group 2 individuals are omitted. We have drawn lines around all individuals within a group, indicating the essentially perpendicular relationship between the major axes of the ellipsoids of dispersion for groups 1 and 3. (After Makurath and Anderson 1973:308.)

point that a bivariate plot of individual specimen scores against the first two discriminant functions clearly indicates that the within-sample pattern of covariation of the two variables is in fact radically different within each of the two species (see figure 6.9). The shape of the ellipsoid of dispersion for the sample of *Gypidula prognostica* (the putative ancestor of *G. coeymanense*) indicates a trend to "long, narrow" shape with increase in size, whereas in the one sample of *G. coeymanense* plotted in like fashion, specimens become "short, wide" with increasing size. It is unclear whether the size-correlated trends within populations are strictly an ontogenetic phenomenon, or a simple covariance of a trend with variation in size, or a mixture of the two. Nonetheless, enough analysis is presented to indicate that the anatomical features of the brachiopods are put together in rather different fashion, suggesting that straightforward, linear modifications of the structures within species is not a model for the transformation of the characters among species.

Another example, drawn from unpublished measurements on some undescribed species of Permian pleurotomariacean gastropods, even more vividly illustrates differences in within-sample patterns of variation between ancestors and descendants. In the following example, patterns of pooled within-species variation along factor analytic axes are compared with among-species patterns. The trends are thus not the same as stratigraphic trends, but nonetheless bear on the issue at hand. In any case, the eight species involved broadly overlap in their stratigraphic distribution. We are grateful to Dr. Roger L. Batten for supplying us with the data for the analysis presented here. Batten (unpublished manuscript) has recognized two species of the conservative gastropod genus *Glabrocingulum* (*Glabrocingulum*) in the Permian of the southwestern United States. In addition, there is one species of the derived subgenus *G.* (*Ananias*) and two species in each of two new subgenera of *Glabrocingulum*. These latter two subgenera are, insofar as is known, endemic to the Permian of this region. They, like *Ananias,* differ from species of *G.* (*Glabrocingulum*) in being more highly spired. These higher-spired taxa differ among themselves largely in terms of differences in whorl shape. In terms of the mathematical parameters of molluscan shell growth (see Raup 1966 for a review), the variation within and among the taxa under study involves predominantly (a) rate of translation of the generating curve along the axis of growth and (b) changes in shape of the generating curve itself. Batten's recognition of taxa is based on a qualitative assessment of these and other features, such as shell ornamentation.

G. (*Glabrocingulum*) is virtually cosmopolitan and its included species occur in Mississippian, Pennsylvanian, and Permian rocks; *G.* (*Ananias*) is found over broad areas in Pennsylvanian and Permian rocks. As stated above, the four remaining species (in the two as yet undescribed subgenera) appear to be Permian endemics of what is now the southwestern United States. Thus, at most, we can construct a cladogram for all species involved, knowing full well that species from other times and places, omitted from the present analy-

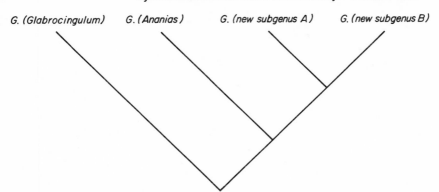

G. (Glabrocingulum)   G. (Ananias)   G. (new subgenus A)   G. (new subgenus B)

Figure 6.10 Cladogram of relationships among the four subgenera of *Glabrocingulum* discussed in the text.

sis, are involved in the phylogenetic history of at least a portion of the group discussed here. But we concur with Batten that the two undescribed subgenera, plus probably the undescribed species of *G. (Ananias)*, form a monophyletic group which in turn is the apomorphic sister-group of the subgenus *G. (Glabrocingulum)*. The cladogram implicit in this statement in shown in figure 6.10.

Thus we have not obtained a precise tree. But we may reasonably inquire whether or not the patterns of variation in the relevant variables within the various species of the plesiomorphic subgenus *G. (Glabrocingulum)* agree with variation in homologous features within and among species in the group of apomorphic subgenera. We have subjected linear measurements on some 187 specimens of these taxa to a factor analysis, where the data were normalized by variables (to give each variable unit variance) and, subsequently, by cases (giving each specimen, in effect, equal size—each specimen being represented by a vector of unit length). Twenty measurements, including spiral angle, were taken on heights and widths of whorls and homologous whorl characters (e.g., selenizone width) on each specimen. (For a similar study, where morphological features and measurements are defined, see Eldredge 1968.)

The first two factors extracted were mirror-imaged (i.e. redundant) and produced a parabolic plot with little taxic discrimination. When specimen scores for factor I were plotted against those of factor III, the four subgenera recognized by Batten were neatly and clearly discriminated. We here show (figure 6.11) a simplified plot of the clusters of each of the subgenera, with individual specimens

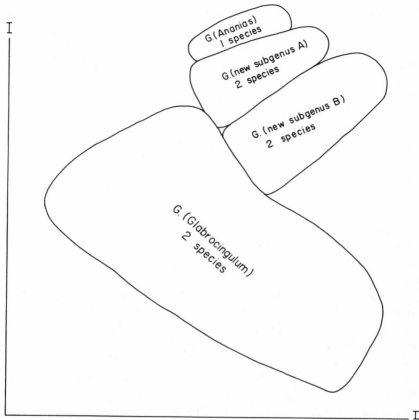

Figure 6.11 Ellipsoids of scores of species within each of the four subgenera of *Glabrocingulum*. (See text for detailed discussion.)

deleted, showing the basic shape of the variation within species (each of the species individually shows the same shape as that shown for the entire subgeneric cluster). Factor I is an axis of variability contrasting relatively narrow, tall shells with relatively short, broad ones. A high score on factor I implies relatively narrower, taller shells. Factor III is an axis of variation pertaining to the upper surface of the whorl in relation to the remainder of the measured morphology. High scores on factor III imply specimens with relatively larger upper whorl surfaces.

With this in mind, and with reference to figure 6.11, the patterns of within- and among-group variation of the *Glabrocingulum* complex can be examined. Both species of G. (*Glabrocingulum*) display a large range of variation, delineating an ellipse of variation whose main axis denotes a relationship of scores of factors I and III with a slope of approximately (-1), i.e., there is a high negative correlation, within these three species, of scores on these two factors. Thus taller, more narrow shells within this group have, if anything, relatively smaller upper whorl faces.

Remaining at the within-species level, it is also apparent that within each of the five species within the three high-spired subgenera there is also a strong correlation between the two factors, but in all of these cases, the relationship is positive: variation in shell height and narrowness is positively correlated with relatively large values for the upper whorl height. Each of the eight individual species displays variation in spire height and narrowness of shell with respect to upper whorl face values, but the nature of this within-species variation differs between the plesiomorphic and apomorphic groups.

At another level, it is also evident that within-species variation in G. (*Glabrocingulum*) is not only greater than within-species variation within the apomorphic group, but also greater than the pooled variation within the entire apomorphic group. More significantly, pooled variation within the derived group is actually in the same direction as that within the G. (*Glabrocingulum*) group. Thus, within the apomorphic group, the subgenera are to be distinguished by differences in whorl height, total height, and narrowness of spire as determined by variation in rate of translation of the generating curve along the axis [just as within the G. (*Glabrocingulum*) group]. It seems to be the case that the *among*-species pattern of variation within the apomorphic group is the same as the *within*-species patterns within the plesiomorphic group—an apparent corroboration of the continuity hypothesis. But the within-species pattern within the apomorphic group happens to be perpendicular to the among-species pattern. Thus the hypothesis is falsified, and in its stead we have corroboration of the decoupling hypothesis, given the validity of the assumptions discussed above.

We draw our third example from a paper written in defense of the continuity hypothesis. Gingerich (1976) has forcefully argued that within-species differences gradually accumulate to produce among-species differences. Especially with regard to size increase (as reflected in the logarithm of the area of the first lower molar in various mammals), trends of increase in size within a lineage as a whole directly reflect within-species trends. These cases where the species are defined on the basis of arbitrarily chosen ranges of values within a plotted continuum, of course, automatically create a situation which confirms the continuity hypothesis. But it is relevant, nonetheless, to note (M. C. McKenna, personal communication, 1978) that in several examples, especially in the graphs [Gingerich 1976, figure 7 (our figure 6.12)] of the data for *Pelycodus ralstoni-trigonodus-jarrovii,* there are a number of "micro trends" towards size reduction of mean size within an overall trend of size increase, possibly providing another example of within-taxon patterns differing from among-taxon patterns.

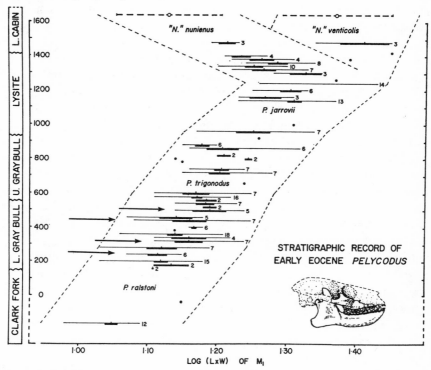

Figure 6.12 Gingerich's diagram of the "evolution of the Early Eocene primate *Pelycodus.*" We have indicated (arrows) the possible "microtrends" in reverse direction relative to the overall trend mentioned in the text. Log (L × W) of M_1 refers to the logarithm of the surface area (product of the length and width of the crown in dorsal view) of the first lower molar. (From Gingerich 1976:16.)

Thus we question seriously the generalization that among-species patterns of net variation arise as large-scale versions of within-species patterns which are, presumably, a function of natural selection. Those who would argue for among-species differences arising as an accumulation of within-species patterns over time cite reversal in direction of the selection vector to cover cases such as the *Pelycodus* example. Like all other forms of selectionist arguments at this level, such a proposition is utterly immune to criticism; it is, rather, a description in dynamic terms (based on a set of assumptions) of data, or refined statistical parameters, plotted on paper. An accurate test can only be made by comparing patterns of variation within and among species that are defined and recognized on independent criteria.

Relationships among the Phenomenological Levels of Evolution: Interconnections

Once the existence of levels, and their mutual "decoupling," is accepted, we must examine the connection between them. There is little evidence that when new species arise from old, their morphological or genetic differences at the outset represent major, sudden shifts (but see Gould 1977a, for the resumption of a contrary view). Rather, available data suggest a spectrum of possibilities, but the overall pattern seems clear enough—there seems to be continuity, or near continuity, in intrinsic attributes among closely related species. Conspicuous morphological gaps within and among monophyletic groups appear to be far more a function of differential extinction than saltatory modifications produced by speciation "events."

This observation in no way contradicts the earlier argument that decoupling among phenomenological levels means that the nature of among-species differences within a monophyletic group is not a function of the accumulation of within-species change over time. When new species arise from old, they most commonly represent a sampling of the ancestral phenotype/genotype—the differences were already there as a subset of the variation within the ancestral species. Some authors (e.g., Mayr 1963; Eldredge and

Gould 1972) have pointed out that if speciation involves relatively small populations isolated on the periphery of the ancestral species' range (where, by definition, edaphic conditions least resemble the norm for the species), selection might be particularly intense, and change might accordingly be much more rapid at the onset of genetic isolation. But such need not necessarily be the case.

Thus speciation—the origin of new reproductive units—does not automatically imply great discontinuity in genetic and phenotypic properties of species. Again we refer the reader to the relationship between selection within populations, on the one hand, and speciation on the other (see chapter 4, p. 121f. and p. 270, this chapter). These ideas form the major theoretical argument that microevolution is decoupled from macroevolution as discrete phenomenological levels. In the allopatric case, natural selection may or may not effect change within isolated populations such that, upon neosympatry, reproductive isolation (via behavioral, anatomical, physiological, or other changes) may or may not eventuate. If reproductive isolation has occurred, it can only be viewed as fortuitous—in no way can selection be said to have acted to create two reproductively isolated taxa from a single ancestral taxon. No benefits could accrue to individuals (the essence of natural selection) from such a phenomenon. Should partial reproductive isolation occur, whereby hybrids produced after neosympatry have reduced fitness, it is consistent with the theory of natural selection to predict that selection will improve reproductive isolation as a necessary consequence of the failure of intergroup offspring to survive. In models of parapatric and sympatric speciation, selection enters earlier to establish reproductive isolation (Bush 1975). Problems with these models are the same as those of envisaging intraspecific, sympatric interdemal selection and will not be discussed further here. The general pattern is that natural selection, as a population dynamic, accounts for differences among ancestral and descendant species, but is blind with respect to the actual creation of new species.

Salthe (1975 in press; personal communication, 1977) has commented on the nature of the connections between the phenomenological levels of evolution. The general nature of such connections is that both lower and higher levels place constraints, or initial and boundary conditions, on any given level. Phenomena at one level thus can feed back and affect phenomena at a lower level. Evolutionary theory to date has insisted that evolution at all levels is essentially a population phenomenon. The view of levels articulated here holds that processes at various levels affect adjacent levels and the vectors can go both ways. The general hypothesis of character displacement provides an excellent example of how phenomena at one level can be "worked out" or "resolved" at a lower level (as noted by Salthe 1975 and personal communication, 1977). We are here concerned with the structure of the hypothesis of character displacement and its relation to levels in the evolutionary process, rather than in the general validity of the hypothesis itself.

The general hypothesis of character displacement is interspecific in nature. It may affect any two species which are competing for resources, or are reproductively similar (see Grant 1972, for a review). Thus character displacement is theoretically expected to be more frequent the more closely related two species are. The hypothesis, in its original form (Brown and Wilson 1956), states that two competing species will be more different in those aspects of the behavior and/or morphology in which the relevant adaptations are manifest in those portions of their ranges where they are in sympatry than where they are in allopatry. Grant (1972) has generalized this hypothesis to read simply that in sympatry, the two species will be either more similar (convergent character displacement) or more different (divergent character displacement) than when in allopatry. In any case, the similarities and differences in sympatry are hypothesized to reflect an adaptive response to the coexistence of the competing species, allowing sympatry to continue. And these adaptive responses in morphology and behavior are effected by natural selection. Thus we have a phenomenon at one level (interspecific competition) resolved (according to the hypothesis) by appropriate within-species (within-population, actually) processes of adaptation via natural selection. One of the outcomes of interspecific competition (according to the hypothesis) is the accommodation by one, or both, species to each other by the modification of the actual local populations involved. The local extinction of a population is another possible outcome. Here, too, resolution of a phenomenon at the interspecific level can be thought of in terms of selection—fitness drops to zero. A third outcome of neosympatry between any two species is a lack of interactive effects at all—which, in a way, can also legitimately (if needlessly) be thought of in terms of selection—as a selection coefficient of zero.

Phenomena at the interspecific level, such as competition, may be resolved in part by processes at a lower level. By the same token, the range of possible reactions at the higher level are clearly bounded, or constrained, by the lower level. Within populations, change via selection is clearly limited by (a) the range of variation, with its myriad underlying controls available, and (b) the limits to which the epigenetic process can be modified at the individual level. The limitation of possibilities for adaptation via selection restrict the range and determine the nature of possible outcomes in terms of new morphologies and behaviors emergent in speciation. The number, nature, and distributions of species existing at any moment, in like manner, determine the nature, and delimit the possibilities of the composition of clades of higher rank at a later time.

Thus the levels are complexly, but directly, interrelated. What goes on at one level simultaneously affects, and is affected by, processes at adjacent levels. But this is by no means to say that the phenomena and processes at these levels are the same throughout, as some aspects of contemporary theory, particularly "synthetic" macroevolutionary theory, are prone to have it. Rather, the levels are at one time decoupled and interacting.

Macroevolutionary Theory: A Restatement of the Problem

Many of the basic questions of evolution remain, whether addressed from a "taxic" or "transformational" point of view, whether phenomenological levels are explicitly considered, and whether species and their origins are included or not. A fundamental task of any evolutionary theory is an elaboration of a coherent theory of how change—genetic, morphologic, and behavioral—is effected. Within-population mechanisms, including natural selection acting on a groundmass of variation, are of great importance in this regard, as we have repeatedly emphasized. However, in the context of macroevolution, there still remains the large subject of rates and directions of such change (i.e., evolutionary vectors) within monophyletic taxa of higher rank. As we have reviewed above, the preponderance of discussion of such macroevolutionary topics denies any phenomenological distinction between within-species evolutionary processes (the mechanics of which are fairly well understood) and among-species processes. In contrast we have adopted the view that within-species processes (specifically, natural selection) indeed effect change in gene frequency, but that smooth extrapolation of considerations of rates and directions of change within species is not an appropriate model for macroevolution, inasmuch as it ignores the problem of the origin of discrete species. We have agreed with Wright and other authors that species origins and patterns of differential species survival are the relevant factors controlling apparent rates and directions of genetic and morphological change among species within a monophyletic taxon of higher catagorical rank. It now remains for us to formulate a general theory of macroevolution in these terms and, subsequently, to consider the testability of its component hypotheses. In formulating such a theory, we are merely recognizing biological principles already well developed in the literature. Moreover, our goal is to show that such a theory can indeed be constructed in testable form; we make no pretensions to formulating a complete or even particularly satisfactory theory of macroevolution.

On a relatively superficial level, the theory of macroevolution states that: (1) All genetic, morphologic, and behavioral change occurring in evolution arises and is maintained by processes at the intraspecific level and below (mutation, natural selection, genetic drift, etc.) but, when seen expressed as inter-taxon differences (interspecific and higher), such change is also the result of the origin and differential survival of species; (2) Net directions of such change among species may be neutral or sustained over a period of time and involve a number of species; and (3) Rates of such change among species depend primarily on the interplay of (a) speciation rates and (b) species extinction rates. In this latter connection, all spindle-diagram phenomena (i.e., patterns of relative diversity through the entire history of a given monophyletic group), including "adaptive radiations," long-term persistence of taxa of low species diversity (and with little net morphologic change, "bradytely"), are to be understood primarily from the perspective of the interplay of speciation and extinction rates.

In elaborating on this skeletal outline of a theory, it is immediately apparent that, just as in microevolutionary theory, there are two components involved. As Raup (1977) has most recently discussed in this connection, Western science continues to seek deterministic explanations for patterns in the natural world. Yet it is quite clear that many of these patterns may arise from pure chance alone. Moreover, what appears to be pure chance at one level may have a deterministic base at another level. Individual point mutations, it is still assumed, each have a specific (deterministic) cause; yet viewed at the population level, what point on a chromosome and what individual within the population will show that mutation is, for all purposes, to be considered random. In addition, mutation is random with respect to the regime of natural selection obtaining within a population at the moment the mutation occurs. Yet the mutation rate within a population as a whole is statistically, at least, a constant, and thus deterministic at the level above the individual. Similarly, speciation is deterministic (in the sense that a set of proximal causes presumably underlies any actual speciation event), yet can be considered random with respect to long-term trends (as Wright pointed out) and, in terms of macroevolution, trends can still be considered deterministic via directed differential species survival ("Wright's rule").

But the probability is at least potentially equal that any such specified patterns are themselves the result of chance. Returning to microevolution for an example, change in gene content and frequency within a population can result from a variety of causes (hence all be "deterministic" in one sense), and yet be either deterministic (natural selection) or random (genetic drift, migration, etc.) in a larger sense. The effect produced, in other words, is best viewed as an accidental product of the process, however deterministic that process might be. Examples include "sampling error" in generational reproductive patterns within populations (genetic drift); origin of reproductive isolation between two groups of populations in allopatry as a sheer accidental byproduct (change in morphology and behavior) effected either by chance (e.g., drift) or selection, but in any case without regard to possibilities of resumption of a pattern of reproductive continuity sometime in the future; at the macroevolutionary level, interspecific trends may reflect some sort of "species selection," but could also arise, one might assume, from chance. The argument in the latter case mirrors that for genetic drift, a concept arising from the troublesome lack of pure and exact correlation between the quality of an individual flourishing within a population and reproductive behavior and "success." Similarly, while extinctions are clearly (and speciations somewhat less so) all caused by some particular concatenation of biotic and abiotic processes, the accidental ("bad luck") component of individual cases of species extinctions is sufficiently apparent to prohibit the elaboration of a purely deterministic theory of macroevolution. To formulate such a theory as purely deterministic would be naive.

Raup (1977) has summarized methods of evaluating such random components of intricate evolutionary patterns. Recognition that there may be "random" components underlying a given pattern has led to some fruitful hypotheses that can be tested (in a statistical sense) in recent years. Raup (1977:63) has recently suggested that species may be regarded as "particles." Assigning probability of extinction of the particle, or production (by splitting) of new particles from old is conceptually in close agreement with the elements of macroevolution discussed here. Raup, Gould and their colleagues (especially Raup et al. 1973; Raup and Gould 1974; Gould et al. 1977) have simulated many patterns of the origin, development and extinction, at the species level, of clades (taxa of rank higher than species). Gould et al. (1977) have compared the shape (summarized as spindle diagrams; see figure 6.13) of simulated clades with those based on actual systematic data and have shown that simulated patterns, involving certain assumptions but in many respects "randomly generated," essentially duplicate the range in "clade shape" seen in systematic data. This work is leading to an evaluation of the "random" component underlying macroevolutionary patterns and, just as significantly, involves the use of actual results of systematic biological investigation—distribution of the components (species) within monophyletic taxa—to test specific hypotheses.

Van Valen (1973) has suggested that, whereas the death of any individual is assumed to have a direct cause (one of many possible), frequency of death of individuals within a population is often so fixed as to allow the estimation of probability of extinction of the population as a whole. He hypothesized that species within a higher taxon may exhibit a characteristic rate of extinction—a statistical generalization comparable to the half-life of a radioactive isotope—and thus the probability of extinction of the entire clade can be assessed. This

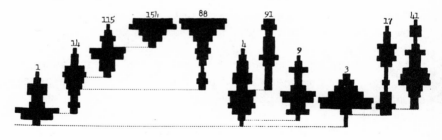

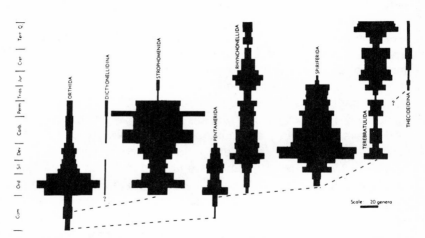

Figure 6.13 Comparison of the spindle diagrams of randomly generated and real clades. Top, spindle diagrams generated by one run of the MBL program at branching and extinction probabilities of 0.1; bottom, spindle diagrams for genera within orders of brachiopods, from the *Treatise on Invertebrate Paleontology*. (From Gould et al., 1977:24.)

approach is again consistent with our view of macroevolution, and offers a means of understanding the random component of macroevolutionary processes.

There are, however, sufficient aspects of patterns of variation within and among taxa of higher categorical rank (e.g., differences in rate and direction of genetic and morphological change and in the production of new species) to warrant a general theory of macroevolution along deterministic lines. Such a deterministic theory would exist as one of two components of unspecified (as yet) importance relative to the random component. The remainder of this chapter is devoted to an elaboration of a deterministic component. It should be borne in mind throughout that the patterns it addresses may also be, in any particular case, randomly generated.

The Deterministic Component
of a Macroevolutionary Theory

As we have outlined, the basic problem of macroevolution is the explanation of genetic, morphological, and behavioral differences among species and taxa of higher rank in terms of the origin and differential survival of species. Rates and directions of such change in particular are a function of the origin, survival, and extinction of species and the interrelationship of these components. Thus any deterministic theory of macroevolutionary processes must be concerned with factors controlling (a) speciation rates, (b) species survival (and its converse, species extinction). It is also clear that theoretical, field, and experimental work pertaining to these subjects is ecological in scope and content.

Ecology, at least that portion of it germane to our present discussion, is primarily concerned with the integration of populations of some finite number of species into a community. The community occupies some specific geographic area (limits may be imprecise and shift through time) and persists for some period of time. Precise species compositions of communities are conventionally theorized to vary both as a function of some developmental process (i.e., a succession of communities, or sere, leading ultimately to a "climax" community) and from oscillations in species composition in communities in equilibrium, reflecting discrete episodes of local extinction of populations, as well as migration of individuals of different species. Thus factors hypothesized to control speciation rates and species persistence and extinction are analyzed and understood at the level of the local ecosystem. At least epistemologically, then, any macroevolutionary theory must consider the ecological controls of speciation and extinction. Whether such a resolution will prove ontologically equally compelling remains to be seen. A further important implication of the relevance of ecological controls on the origin, persistence, and extinction of species is that the concept of adaptation can be brought back into our consideration of macroevolutionary theory.

There is a set of relationships which describes in general terms the role of ecological theory in tying together such disparate strands of evolutionary phenomena as adaptation, on the one hand, and species diversity (encompassig origin, persistence, and extinction) on the other. The relationship, stylized as an equation in figure 6.14, hinges on the fact that both species diversity and the morphology of individuals are related to the occupation and exploitation of ecological niches. Therefore niche theory should act as an interactive catalyst in linking these two topics in evolutionary theory.

In the first statement of figure 6.14, morphology is implicated with niche theory. All this says is that organisms are adapted to the ecological niches they occupy (a tautology) and that such adaptations are expressed in the intrinsic properties of individuals comprising the local population occupying a given niche. (If this latter point is not exactly a tautology, it is nearly so.) In other words, traits of organisms are either directly related to the occupation and exploitation of the population's niche, or, at the very least, neutral (irrelevant) with respect to the occupation of the niche. Traits which appear unrelated or superfluous to the exploitation of a particular niche abound in all species (e.g., the presence of five arms, not four or six, in ophiuroids

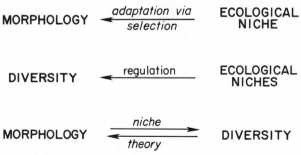

Figure 6.14 Three stylized "equations." The upper relates morphology to ecological niches effected by adaptation via natural selection; the middle statement recognizes the control over species diversity exerted by niches (number and breadth). The two upper statements are therefore related, with niche theory serving as the common denominator.

scuttling over the sea bottom). While some of these traits may, in fact, represent adaptively neutral characters arising by pure chance in speciation, most such examples (such as the ophiuroids) represent characters with a higher level of distribution (all ophiuroids possess five arms; all but the most primitive echinoderms known are pentaradial). In general, as an axiom of the evolutionary system, synapomorphies—at whatever hierarchical (taxonomic) level—represent evolutionary novelties. Evolutionary novelties, according to the highly corroborated theory of microevolution, generally arise as an outcome of natural selection on available variation which maximizes adaptation of a population to the particular exigencies of a local environment—the population's niche. Thus there is a firm conceptual connection between morphology and ecological niches, a connection based on adaptation via selection.

The second line of figure 6.14 relates diversity to niche theory. Regulation of species diversity has become one of the more active areas of ecological and evolutionary research during the past 20 years—so much so that the two books written by paleontologists in the United States since 1953 (the year of publication of Simpson's *Major Features of Evolution*) which include the word "evolution" in the title are primarily concerned with the origin, maintenance, and degradation of taxic diversity (within ecosystems, not necessarily within monophyletic groups as discussed here). Both Valentine's (1973) *Evolutionary Paleoecology of the Marine Biosphere* and Boucot's (1975) *Evolution and Extinction Rate Controls* come close to treating evolution strictly as a matter of diversity regulation. Such a shift in emphasis since Simpson's work reflects, at least in part, the large amount of effort in theoretical ecology since World War II devoted to the ecological regulation of species diversity.

Species diversity reflects an interplay between rates of appearance (by migration or evolution) and disappearance of species, and ecologists have been ineluctably led to consider these issues in terms of niches. Considerations of the regulation of diversity have involved correlations between geographic area and diversity (equilibrium biogeography, started by MacArthur and Wilson 1967), correlation between latitudinal gradients and diversity, correlation between water depth and diversity in marine environments, and the effects of ecological and geological time on diversity. Underlying each of these formulations is the concept of niche number—again, as a tautology, the number of species present is equal to the number of niches realized (occupied and exploited in a given area). This is true even if the analysis is statistical and merely seeks to correlate species diversity with, say, geographic area. Thus any consideration of the origin and maintenance of diversity must consider problems of ecological niches themselves, including relative niche breadth, changes in niche breadth, the effects of interspecific competition, resource partitioning, and the like.

Utilization of resources and problems of relative niche breadth simultaneously lie at the heart of the controls of species diversity, on the one hand, and on the other hand, are the bases of the adaptational (functional and behavioral) characteristics of any given species. Thus we can connect the first two lines of figure 6.14 into a single "equation" as given in the third line of the figure. Niche theory provides the nexus between considerations of adaptation and considerations of the origin, maintenance, and degradation of species diversity.

It is not our purpose, in a general book on phylogeny reconstruction and aspects of macroevolutionary theory, to discuss in detail contemporary ecological theory pertaining to niches, either from the standpoint of diversity or from the standpoint of functional anatomical analysis of adaptations (niche utilization). Such theory in any case cannot be tested with reference to branching diagrams in systematics, the central concern of this book. The purpose of this section has simply been to suggest the nature of the dynamic process involved in macroevolution, as well as the nature of the connection between two areas (i.e., adaptation and species diversity) conceptually divorced earlier in this chapter.

Hypotheses of macroevolutionary processes involve speciation and species persistence and extinction, and their interrelationships. Subordinate hypotheses pertaining to these processes belong to the realm of ecology and ecological theory. Such theory is focused on interspecific interactions (especially competition) and resultant effects in terms of niche width. Relative niche width is, in turn, related to the concepts of relative eurytopy-stenotopy. Originally defined in terms of habitat breadth, the terms eurytopy and stenotopy have become more conventionally used to refer to relative breadth of tolerance to specifiable parameters of the physical and biotic environment, or the relative breadth of ability to exploit specifiable parameters of the resource space. The terms as used in ecology have rough equivalents in the evolutionary expressions "broadly adapted" and "narrowly adapted" (i.e., when those expressions are applied to single species) and these latter terms are themselves near-synonyms of "generalists" and "specialists." These various pairs of terms are not strictly synonymous, but are sufficiently close to permit a general discussion of the nature of the relationship between these variables and others encountered in macroevolution.

Theory and Information in Macroevolutionary Analysis

The systematist attempting to confront the issues of macroevolution has the following sorts of data (in the form of well-corroborated hypotheses) to work with: (a) a theory of relationships among species within a monophyletic taxon of higher categorical rank; (b) the distributions of the component species in space and, where possible, in time (i.e., fossil evidence); and (c) an evaluation of the relative degree of apomorphy of each specifiable morphological feature in each of the species considered. This last system of hypotheses arises from the very analysis upon which the theory of interspecific relationships is based. When the relative distributions of homologous features are viewed from the standpoint of evolutionary novelties, a conceptual link between adaptive specialization and relative apomorphy, vs. adaptive generalization and relative plesiomorphy emerges. The relationship holds only for comparisons of homologous traits among species where the one (or more) relatively apomorphous condition may be hypothesized to constitute a functional specialization. With respect to that particular anatomical specialization—assumed or hypothesized to relate to the tolerance for, or utilization of, some specifiable component of a species' niche—species lacking the apomorphous condition (i.e., retaining the more plesiomorphous state) are hypothesized to be more generalized with respect to the homologous niche parameter.

In plainer language, we simply reiterate the oft-noted correlation between anatomical specialization on the one hand, and behavioral and functional, hence adaptive, specialization on the other, and this with a narrowing of range (with respect to a certain parameter) of a niche. Thus, with respect to comparison of niche width among species related in some fashion on a cladogram, it may be hypothesized and tested for individual characters, that relatively more apomorphous species occupy narrower niches, at least in terms of niche parameters associated with those particular anatomical and behavioral attributes. We suggest that this general sort of correlation between evolutionary (anatomical and behavioral) specialization and niche width among a series of closely related species, is a fairly well-corroborated generalization already; but we point out that field ecological analysis coupled with functional anatomical analysis can test the hypothesis in any given instance. All that is needed at the outset is a well-corroborated hypothesis of relationships among the species involved. The correlation between anatomical and behavioral specialization on the one hand, and narrowing of width of a particular niche on the other, and the converse (ecological generalists with relatively plesiomorphous morphologies) arises as a generalization from such work as Bock's (1970, 1972) on the Hawaiian Drepanididae, and Lack's (1947) analysis of Darwin's finches on the Galapagos Islands. But we repeat that such a blanket generalization need not be accepted for our argument; rather we simply point to the conceptual link between morphology and niche width and further suggest that in any specific case (involving Recent species, at least), hypotheses of this form are directly testable.

The original definitions of "eurytopy" and "stenotopy" referred to habitat breadth and were used in ecological geography almost as synonyms for "widespread" and "narrowly distributed." These latter terms, of course, further suggest the contrasting set of terms "cosmopolitan" and "endemic," terms generally used as summary descriptors of relative geographic spread, devoid of any implication of ecological specialization or generality. We do not mean to suggest that eurytopic species cannot be endemic to small areas or that stenotopes cannot be widespread. Nonetheless, as a rule and within a monophyletic group, cosmopolitan species tend to be eurytopes (in the sense of being relative ecological generalists) whereas relatively more restricted species tend to be stenotopes. Again, the reader need not accept this generalization as ironclad; for any given case, the hypothesis can be tested directly, and all that is needed at the outset, is, once again, a well-corroborated theory of relationships among the species involved.

While on the subject of the correlation between species distributions and relative niche breadth, we note the generalization—going back in the literature at least as far as Williams (1910)—that narrowly adapted species (as judged by lack of within-species variability or presence of autapomorphies) not only tend to occur in relatively restricted geographic regions, but also tend to persist for shorter periods of time in the stratigraphic record. This suggests, of course, that specialized species are relatively more prone to extinction, a long-standing suspicion in evolutionary theory. The problem with this particular generalization is that it is extremely difficult to test; relative duration in the stratigraphic column may be highly corroborated (in spite of the sampling problems inherent in studying the fossil record), but evaluation of relative eurytopy/stenotopy must be based on ancillary, secondary correlations, which themselves are untestable and possibly false for any particular case. In fact, the generalization may not hold: relative lack of variability may indeed correlate well with stenotopy within a group of related species, but then again it may not. Evaluation of relative stenotopy and eurytopy among fossils hinges on the assumption that such a relationship pertains, or is based on geographic distributional data (i.e., range of occurrence in reconstructed paleoenvironments). In either case, relative eurytopy/stenotopy among fossil species is not susceptible to testing and thus the generalization about the correlation between such ecological strategies and temporal duration is itself not susceptible to testing in any particular case.

However, distributions of species in time and space can be compared directly with a matrix of relatively primitive/derived morphological features. Thus the hypothesis that relatively plesiomorphous species within a monophyletic group tend to be more widespread geographically or to persist longer (stratigraphically) can be easily tested. All that is required is a prior, highly corroborated theory of relationships among the species, and the relevant distributional data. The only further requirement is that the theory of relationships cannot itself be based, in whole or in part, on the distributional data.

There is thus a conceptual, and for the most part, for specific cases, testable set of three interrelationships between three classes of variables: (a) relative niche breadth, (b) relative degree of apomorphy, and (c) distribution of species in space and time, all in the context of well-corroborated hypotheses of relationship among the species involved. If these three sets of variables are arguably, but testably, interrelated, it still remains to be shown how they might be implicated in the control of speciation, species persistence, and species extinction.

At this juncture, we reiterate our acknowledgment of "random" factors; a new species may appear as the accidental by-product of change in the physical geography of the area, or a portion thereof, of its ancestor. A species might last longer than its sister in the next valley for purely accidental reasons, the converse of the observation that many extinctions are unlucky accidents. Entire ecosystems can be degraded relatively quickly (less than a million years, to judge from such "events" in the Upper Devonian, Upper Permian, Upper Triassic, Upper Cretaceous, and Pleistocene, among others). In such cases, involving thousands of species in many unrelated clades, an environmental event (literally a catastrophe) occurs that is utterly accidental with respect to individual species adaptations and particularly random with respect to the interspecific interactions and resultant patterns of differential species survival resulting from deterministic macroevolutionary processes within monophyletic groups.

With this in mind, it is now relevant to consider the relationship the three sets of variables discussed above might theoretically have to patterns of differential species survival—macroevolution. If species diversity is, at base, a function of niche width, a general set of predictions about distributions in space and time of relatively apomorphous/plesiomorphous species emerges. In general, within a single monophyletic group, highly speciose groups (at any one given time, throughout the entire collective geographic range of the species-group) should be (in terms of each component species) relatively narrow-niched, narrowly dispersed geographically, and relatively apomorphous (including rather autapomorphous) (see Bretsky and Lorenz 1969, for a similar list of correlated variables). Individual species within less speciose sister-groups should prove to be (a) relatively plesiomorphous (b) more widespread geographically, and (c) broader-niched. We cite Fryer and Iles (1969) and Greenwood's (1974) analyses of species flocks of cichlid fishes in east Africa, if not as a conclusive test of the generality, at least as an example that fails to falsify it. The cichlid genus *Tilapia* is relatively nonspeciose (depauperate), and its included species tend to be relatively broad-niched and geographically widespread (within a given lake or drainage system) than species of the related, highly speciose genus *Haplochromis*.

There is one important objection to the correlation between niche width and diversity as a means of approaching macroevolutionary theory: ecological theories of diversity, including niche width and interspecific interactions (especially competition), pertain to all species within a geographic area, not to all component species of a monophyletic group, regardless of their occurrence (i.e., in allopatry or sympatry). Thus we conclude that if such theory be relevant to macroevolutionary phenomena involving monophyletic groups, then the discussion is limited to interspecific interactions among sympatric and parapatric species within the group—a condition already clearly recognized by Bock (1970, 1972) in his analysis of the evolution of Drepanididae.

What, then, are the ultimate conceptual links between niche breadth on the one hand, and control of species diversity within monophyletic groups (speciations, species persistence, and extinctions) on the other? Although detailed ecological theory is both inappropriate for this book and, in any case, outside the bounds of our expertise, the basic nature of this relationship does seem fairly clear, and is susceptible, of course, to further examination. We have already noted the healthy disagreement in contemporary ecology over the issues of the origin, maintenance, and degradation of diversity patterns. One of us (Eldredge 1979b) has hypothesized that speciation rates themselves are, in an important sense, a function of the "niche strategy" of individual species within a monophyletic group. The hypothesis states that, within a given group eurytopic species react differently to interspecific competition than do stenotopes; eurytopes tend to react to interspecific competition by mutual exclusion, whereas stenotopes more commonly react to such competition by further subdivision of resource space (niches are further narrowed). Hence, within a monophyletic group, if there is a discernible spectrum of stenotopy and eurytopy, there are frequently many stenotopic species and rather fewer eurytopic species. The hypothesis is suggested by actual patterns of distribution in nature: within monophyletic groups relatively eurytopic species tend to be allopatric with respect to one another (or "vicariant") and far-flung, whereas congeneric stenoptopes are more likely to occur sympatrically.

There are problems with this hypothesis. Why, for instance, do eury-topic species comprising a relatively eurytopic species lineage, *remain* eurytopic? In any case, the hypothesis requires a great deal more testing (and mathematical investigation) than we can give it here. We also point out that the hypothesis in essence constitutes an argument about dampening controls on speciation rates, though the converse (that stenotopy actually *causes* high rates of speciation) is not held to be true. We include it here as an example of the kind of conceptual link between rates of speciation and niche width required in macroevolutionary theory.

We have already briefly mentioned the supposed connection between niche width and extinction (species persistence being the opposite of extinction, the two are treated as the same problem here). According to conventional wisdom, more broadly adapted organisms (eurytopes) are expected a priori to be able to survive unpredictable and large-magnitude environmental disruptions because they exhibit (by definition) broader physiological tolerance ranges. The hypothesis is intuitively appealing and seems well corroborated in mathematical and experimental treatments given it by ecologists (see Bretsky and Lorenz 1970). Insofar as there is a deterministic element to relative rates of extinction within monophyletic groups, patterns of relative niche width are at least implicated if not exclusively the controlling factors involved.

Macroevolution: Hypotheses and Some Basic Patterns

We have adopted the general assumption that macroevolution is basically a phenomenon of differential species survival. If the line of argument linking adaptation and species diversity through utilization of species' niche widths is basically correct, macroevolutionary patterns (to the extent that they are deterministic at all) can be investigated by formulating predictions about the nature and distributions of individual species sampled in the monophyletic group under study. The general procedure involves three steps:

1. Perform phylogenetic analysis, arriving at a well-corroborated cladogram of relationships among all species sampled

(evolutionary analysis of patently non-monophyletic, including "gradal," groups, under these terms is obviously meaningless).

2. Tabulate the pattern of species diversity.

For Recent organisms, the pattern is simply a tabulation of all known species. For groups known in whole or in part from the fossil record, a time-averaged appraisal of the "standing crop" of known species within the shortest recognizable interval of time is used.

3. Depending upon the characteristics of the pattern seen in (2), specific predictions are made about the component species sampled.

The predictions deal with the geographic and stratigraphic distribution of each species in conjunction with the distribution of apomorphous and plesiomorphous characters and, especially with Recent species, relative degree of eurytopy and stenotopy. If the predictions do not agree with the observed (i.e., hypothesized but corroborated) attributes of the component species, the general set of predictions about diversity controls is falsified. Should successive examples also be falsified, the basic assumption—that diversity and adaptation are a function of modes of ecological niche occupation and exploitation—can be seriously questioned. Testing such complex propositions is always difficult. Potential error in tabulating distribution patterns and in characterizing stenotopy vs. eurytopy is so great that we suggest rejection of a macroevolutionary hypothesis cast in these terms only if 6 percent or more of the putative cases fail to substantiate the predictions. We are aware that, even if our proposed rejection of the conventional extrapolation of microevolutionary processes to the level of macroevolutionary phenomena is accepted, our proposed reorganization of macroevolutionary theory may not be successful. But at least we have identified a means whereby our alternative theory can be directly evaluated, criticized, and perhaps rejected. And perhaps it may eventually emerge well-corroborated.

The specific kinds of diversity patterns addressed below deal with both fossils and Recent organisms. Both are important, for different reasons. Patterns of diversity in the fossil record (which can be graphed in a variety of ways, including the familiar spindle diagram) display a sampling of actual patterns in true evolutionary time. Only with a fossil record can changes in diversity patterns within a group be tabulated. And part of the complex of predictions important in assessing controls of the rates of both speciation and extinction require accurate evaluation of at least relative (absolute would even be better) duration of species in time. Only the fossil record can supply such information. On the other hand, evaluating relative eurytopy and stenotopy—a difficult task at best—can only be done with any degree of credibility with Recent organisms. And the Recent biota is, after all, a direct product of differential species survival from the Pleistocene epoch. Thus analyses, such as Bock's (1970, 1972) on the "adaptive radiation" of Drepanididae, are by definition valid and contain much functional anatomical appreciation of niche utilization not basically possible to secure with fossils.

Table 6.1 is a classification of macroevolutionary patterns according to (a) criteria of recognition (diversity pattern, ratio of extinction to speciation, and absolute rate of speciation), (b) hypothetical nature of pattern of species selection, and (c) predictions about the nature of component species for each pattern. Other patterns could no doubt be listed; the classification is intended to be illustrative rather than exhaustive. More important, other theories of speciation and extinction controls will produce other predictions. We aim in this section to show how macroevolutionary theory can be hypothetico-deductive under the general notion of differential species survival and not necessarily to establish a specific version of such theory. We shall now discuss each macroevolutionary pattern in the order presented in table 6.1.

Adaptive Radiations

The expression "adaptive radiation" is itself a value-laden description, hinting at the causes underlying the particular pattern. A more neutral definition of the pattern is simply a relatively rapid proliferation of species of a monophyletic group—"relatively rapid" contrasts the rate of proliferation of species within the group both before and after the radiation begins. Although such radiations may occur toward the beginning of the history of a clade, such need not be the case.

The adjective "adaptive" appears to be appended to the general term for this pattern because the proliferation is felt to result directly from the sudden availability of a new opportunity. The new opportunity might simply be new habitat space (e.g., an island unoccupied by ecologically similar organisms), or new "adaptive

Table 6.1 Macroevolutionary Patterns: Criteria for Recognition, Mode of Species Selection, and Predictions Concerning Component Species

Macroevolutionary pattern	Criteria					Predictions		
	Diversity pattern	Sp/ext ratio[a]	Sp rate[a]	Spp sel[a]	Ext rate[a]	Eury/steno[a]	Geog dist[a]	Apo/plesio[a]
Adaptive radiation	Many closely related species arising in a short period of time	$\gg 1$, later, ~ 1	High	Disruptive	High	Mostly stenotopes; much sympatry	Narrow	Many autapomorphies; plesiomorphies relatively few
Arrested evolution	Low diversity for long periods of time	~ 1	Low	Neutral	Low	Eurytopes; little sympatry	Wide	Much retention of plesiomorphies; autapomorphies rare
Steady state	Moderate diversity for long periods; occasional adjustments in equilibrium value	~ 1	Moderate to high	Neutral or centripetal	Moderate to rapid	Mixed	Varied	Some subgroups with autapomorphies; some retention of plesiomorphies
Trends	Varied, usually low diversity	~ 1	Moderate to high	Directional	Moderate to rapid	Stenotopy dominant	Narrow	Clinal, progressive replacement of successively more apomorphic states

[a]Abbreviations: Sp/ext ratio, ratio of speciation and extinction; sp rate, speciation rate; spp sel, mode of species selection; ext rate, extinction rate; eury/steno, degree to which component species are eurytopic or stenotopic; geog dist, relative geographic distribution of component species; apo/plesio, relative degree of apomorphy and plesiomorphy of component species.

space" (e.g., invasion of the terrestrial environment by aquatic organisms, or the invention of a new behavioral and/or anatomical complex which so improves utilization of the ecological space that the way is opened up for the existence of a host of variations on a theme—one way of dividing up the new or "redefined" resource space). Simpson calls the latter phenomenon "key innovations."[11]

Whatever the mechanism involved, as pure pattern, adaptive radiations consist of many closely related species occupying a discrete geographic region. Sympatry, or near sympatry, tends to be high. Speciation rate, of course, is high and exceeds extinction rate. As a non-testable descriptor of the pattern, species selection is disruptive—many variations on a morphological theme being produced (perhaps as in Bock's vision of divergent character displacement resulting from interspecific competition among closely related species of Drepanididae).

From this pattern, a set of specific hypotheses about the component species can be adduced: (1) Relatively stenotopic species should greatly outnumber eurytopes; thus (2) Extinction rates should be fairly high (i.e., component species should be fairly short-lived); (3) Each species should be restricted to a relatively small portion of the geographic range occupied collectively by the entire group; and (4) There should be a high percentage of species marked with a number of autapomorphies, with synapomorphies among subgroups of species relatively less common.

We now take a pattern conforming to the descriptive criteria of a radiation, characterize the pattern, and apply the set of predictions. The pattern stems from the original research of one of us (N.E.) and is at this date only partially published, in the systematic literature. Moreover, we state at the outset that the predictions are confirmed. We present this analysis for its heuristic value, to show that macroevolutionary hypotheses involving species origins and extinctions can be tested. Other similarly patterned data may well refute some or all of the predictions. The point is that such falsification can

11. A related concept, that of "adaptive zones" (see p. 261), seems to be to taxa of higher rank as the concept of niches is to species. Since such "higher" taxa do not exist as ecological entities, they can hardly invade or occupy anything. If there is anything to the concept of the "adaptive zone," it is as a summation of all the realized niches of all the component species within the larger monophyletic taxon; the relation of this concept to evolutionary theory is unclear.

occur, not that the hypothesis of adaptive radiations presented here is correct in any or all of its details.

The test case involves the Calmoniidae, a family of acastid trilobites (superfamily Acastacea; see Eldredge, 1979a, for a brief discussion of higher-level systematics of these trilobites). The Calmoniidae were endemic to the Malvinokaffric Faunal Province of the Upper Silurian and Lower and Middle Devonian. The faunal province encompassed the marine environment south of 60° south latitude, and included an Andean component (southern Peru, Bolivia, and northern Argentina), the Amazon Basin of Brazil, the Falkland Islands, and South Africa. Ghana and Antarctica may also have belonged to the province. Geophysical evidence places the South Pole just north of Cape Town in the Lower Devonian. All of these continental areas were united into the supercontinent Gondwana during this time. The total span of time during which this clade persisted was at least 20 million years. However, the record is brief on all platform areas, and it is only within the Andean sequence that anything like a continuous stratigraphic record is available, with much more of the 20 million years represented in some local sections up to 10,000 feet in thickness. Stratigraphic correlation between areas is still quite rudimentary, and the amount of time missing from the record between the Upper Silurian and the lowermost Devonian beds is not as yet even roughly known. Despite the patchy distribution of rocks and fossils, enough of a pattern remains to indicate a radiation.

The Calmoniidae consist of some 29 known genera and subgenera, and at least 60 species, according to a recent tabulation (Eldredge and Ormiston 1979). If this seems a moderate diversity of twenty million years or so, it is to be recalled that (1) Only some fraction of existing species have been fossilized, sampled, and described and (2) Nonetheless, compared with most other known trilobite groups over comparable spans of time, diversity is high indeed (there are, in addition, many other noncalmoniid trilobites in the fauna). Thus the pace of evolution, at least by trilobite standards, was rather high for Calmoniidae.

In figure 6.15 we present a cladogram of relationships among all described genera and subgenera. The cladogram can hardly be said to be "highly" corroborated; it is defended in outline in Eldredge and Branisa (1980) and more rigorously and completely in

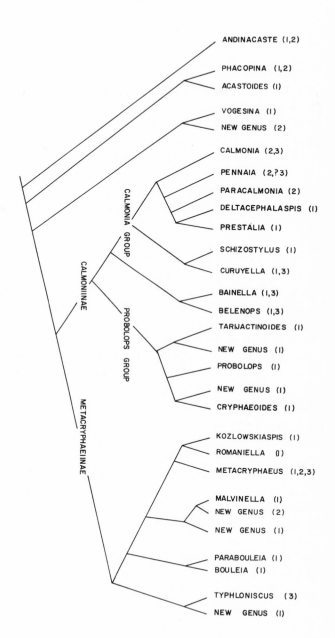

Figure 6.15 Cladogram of relationships among calmoniid genera and subgenera, with major groups labeled. Numbers beside taxon names refer to the geographic occurrence of component species: (1) Andean South America; (2) non-Andean South America; (3) Falkland Islands and South Africa.

Eldredge (unpublished manuscript). The range of anatomical variation within the Calmoniidae is unusually great. The pattern of relationship shown on the cladogram indicates that the variability is parcelled out in a regular fashion: there are (a) a few calmoniids so plesiomorphic as to resemble most closely their more primitive acastid relatives. (Acastids are cosmopolitan in the Silurian and are known from Australia and the Northern Hemisphere in the Devonian.) The remaining calmoniids are divided into two subfamilies, the (b) Calmoniinae, containing two related subgroups, in which there is an array of different forms united by a distinctive set of features, and (c) the Metacryphaeinae, members of which, while united in some features, are greatly different one from another and reminiscent of other, nonacastid groups of trilobites. The Metacryphaeinae thus fulfill the standard conception of an adaptive radiation. Details of the composition of these groups, including the anatomical peculiarities of the component species and genera, are given in Eldredge (unpublished manuscript).

We now compare the set of predictions with properties (observed or hypothesized) of the component species. The first criterion, that stenotopes ought to outnumber eurytopes, cannot be examined directly. Indirectly, the restriction of most species to specific sedimentary environments (see Eldredge and Ormiston 1979), as well as to highly localized areas, agrees with, but does not critically test, the prediction. We also predict high rates of extinction (relatively short stratigraphic ranges), a high degree of (sub)regional endemism, and a characteristic preponderant development of autapomorohic features as anatomical markers of specific-level adaptations. Figure 6.16 is a chart of the approxiate stratigraphic ranges of all known calmoniid species in Bolivia.[12] The plesiomorphic genera *Acastoides* and *Vogesina* each have long-ranging species, and each genus persists through most of the total time span. Only species of *Schizostylus, Metacryphaeus, Malvinella,* and *Bouleia* exhibit comparable ranges. Apart from the two species of the

---

12. Correlation of the Andean sequence with other regions in the Malvinokaffric Province is insufficiently precise to incorporate non-Andean calmoniids on this chart. Inasmuch as 23 of the 29 genera and subgenera of Calmonidae (i.e., 79 percent) occur in the Andean region, the chart of figure 6.16 should effectively mirror the gross pattern of temporal variation among species of the Calmoniidae. The chart is stylized; e.g., the concordant ranges within the *Scaphiocoelia* zone reflect isolated presence within the zone, not necessarily concordant appearance and extinction events.

Silurian plesiomorphic taxon *Andinacaste,* the beginning of the radiation appears in the *Scaphiocoelia* beds, with 2 of the 14 calmoniid species plesiomorphic acastidlike forms, the rest being rather highly derived. The *Calmonia* group presents a rather odd mixture of primitive and derived forms: the group as a whole retains the plesiomorphic, typically acastid construction of the central region of the head (glabella), and at the same time, each subgenus is marked with a striking set of autapomorphies involving development of unusual spinescence and other features. This group undergoes its own "radiation," and only one genus (*Schizostylus*) has surviving species in younger horizons. Thus, in some respects, the more primitive members of the Calmoniidae, the *Calmonia* group of the Calmoniinae, occur in the earliest horizons—yet each species is, as well, highly autapomorphic, presumably reflecting specialization. Each species is short-ranged and highly localized. There is no sympatry among con(sub)generic species (according to the taxonomic scheme adopted prior to this analysis), but there is extensive sympatry among species of these closely related genera.[13]

The *Probolops* group is composed of five known species, each quite autapomorphic and each with a very short known stratigraphic range. The group is entirely confined to the Andean region. Nonetheless, the group is present throughout the duration of Malvinokaffric provincial time. We return to the *Probolops* group, as an example of trends, below.

The anatomically highly varied *Metacryphaeus* group is represented in the *Scaphiocoelia* Zone by two species of *Romaniella,* the apomorphic sister-group of the genus *Kozlowskiaspis,* and by a species of *Parabouleia.* Originally described (Eldredge 1972b) as the primitive ancestor of the highly apomorphic *Bouleia,* additional material has shown that, in addition to the relatively plesiomorphic glabellar features, *Parabouleia* possesses a highly derived, stalked lensless "eye," whereas its supposed descendant, *Bouleia,* retains normal calmoniid eyes. The group dominates the bulk of the middle and upper sediments of the Devonian sequence in Bolivia, and shows a mixture of short- and long-ranging species. The group is

---

13. Given the fact that the Malvinokaffric Province occupied a relatively small area, there is an extremely high degree of endemism within it. Geographic ranges are given on the cladogram (figure 6.15); only *Metacryphaeus* occurs in all three recognized subregions of the province. The biogeography of these taxa is discussed extensively in Eldredge and Ormiston (1979).

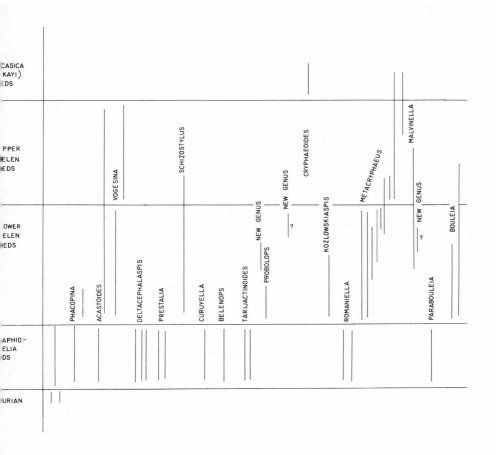

Figure 6.16 The approximate relative stratigraphic ranges of known species of calmoniid trilobites of Bolivia. (See text.)

highly divergent, there being no discernible spectrum of primitive-derived features within the cladogram. (We mention this group again in another connection below.)

There is, clearly, a great deal more to be said about these trilobites; the analysis, for the present purposes, must be rather skimpy. Enough of the pattern has been presented, however, to indicate that an adaptive radiation, when viewed as a pattern of species origins and differential extinctions, is a complex affair, with monophyletic subgroups developing their own patterns. We should, at this point, reiterate that it is as yet unclear how deterministic components of these patterns may be discriminated from random components (see p. 298 for discussion). No doubt many of these patterns are due to chance alone; still more may be attributable to sampling error. It does not do to be overly optimistic in framing deterministic just-so stories about such patterns. But the predictions about the species components which arise from the "differential species survival" approach to macroevolution, hold true in a general way. More importantly, they might not have held true, and may not in subsequent analyses of these and other sets of data. We believe that other data pertaining to apparent "adaptive radiations" should be examined from this perspective.

Arrested Evolution

Arrested evolution [Simpson's (1944) "bradytely"] has been traditionally defined as little or no morphological change within species of a monophyletic lineage. Commonly, cases involve at least 100 million years, and frequently more, where a fossil is compared directly with a nearly identical living representative. So-called "living fossils," moreover, tend to be members of plesiomorphic clades— oppossums (metatherian mammals) vs. eutherian mammals, lingulid (inarticulate brachiopods) vs. other brachiopods, *Neopilina* (a monoplacophoran mollusc) vs. other Mollusca, and so forth. (We admit that these "plesiomorphic clades" may not all be monophyletic, though none is blatantly a nonmonophyletic, not-A group). Simpson (1953:319 ff. and 303 ff.) presents an excellent discussion of this general phenomenon from this point of view.

Another general feature of bradytelic lineages, less frequently discussed, is their low species diversity.[14] Given the possibility that it is the constant low diversity through time of bradytelic lineages (commonly, but not necessarily, after an initial "burst," or radiation) that has controlled the low observed rate of morphological change, arrested evolution can be characterized (see table 6.1) as low diversity for over, say, 80–90 percent of the time of a clade's persistence, a nearly equal ratio of speciation and extinction, and a low rate of speciation. These are commonly observed patterns, as for example in dipnoan and xiphosuran phylogeny, as graphed in Eldredge (1979b).

The mode of species selection would appear to be "neutral"—that is to say, one species is much like another throughout the geographic range and the geologic history of the group. The constituent species are, hypothetically at least, nearly interchangeable. This consideration arises from the testable predictions: species within bradytelic lineages are eurytopic, geographically widespread, display long stratigraphic ranges (rate of extinction is low), and rarely develop autapomorphic anatomical specializations, but instead retain and pass along plesiomorphic anatomical features. Interspecific morphologic differences are subtle—there is a monotony about the collective morphologies of these groups, which of course initially inspired the question of arrested evolution. Eldredge (1975 and especially 1979b) has considered the conceptual link between these parameters at length and has claimed that the hypothesis of bradytely stated above, when tested against actual examples, has yet to be refuted. The reader is referred to these papers and to Stanley (1975) for a more detailed discussion of the phenomenon from the standpoint of "species selection."

Steady State

The next macroevolutionary pattern in the list of table 6.1 is the most important—at least if "most common" suffices as a criterion of high

14. Eldredge (1979b) has discussed the chicken-and-egg dilemma inherent in this problem: do bradytelic lineages display low diversity because there is too little morphological transformation to allow a systematist to "see" change and thus recognize and name more species or are speciation rates so low in these lineages that little morphological change can occur? The former possibility keeps the issue safely within standard transformational lines of argumentation, whereas the latter expands the question into the area of interspecific macroevolutionary phenomena as here construed.

importance. Simply labelled "steady state," we refer to the situation of long periods of equilibrium diversity shown by the bulk of groups of organisms represented in the fossil record. Major changes in diversity typically are correlated with similar events in other, unrelated monophyletic groups—particularly episodes of widespread extinction, followed in some instances by a rediversification which may or may not bring diversity rapidly back to a level even higher than the prevailing equilibrium level prior to extinction. Whatever its specific manifestation, steady state is "normality." It is also macroevolution (though rarely so considered) as it clearly represents a pattern of speciation and differential species survival, an equilibrium interplay whereby the ratio of speciation to extinction is, on the average, about one, and rates of speciation can be low, but are usually moderate to rapid.

The mode of species selection is hypothesized to be either neutral or "centripetal," where radically new morphologies, when appearing, do not lead to new subclades, hence little substantial divergence (anatomically speaking) occurs. The reason this general area of "normal macroevolution" is little discussed is that the basic characterization of its essential pattern (moderate to high speciation rates, neutral or centripetal species selection) is rather messy and imprecise, as are the attendant predictions. These are that there should be (a) a mixture of eurytopic and stenotopic ecological strategies; (b) a range from high endemism to virtual cosmopolitanism in geographic occurrences of individual species; (c) a wide range in the duration of stratigraphic occurrences of individual species; and (d) a mixture of specialized and generalized (relatively apomorphic and plesiomorphic) forms. Furthermore, these variables might be apportioned in a nonrandom way among monophyletic subgroups (e.g., some species-groups may be more eurytopic and plesiomorphic, etc. than their sister groups), but the variables might also be "normally" distributed such that, for instance, there may be a spectrum of stratigraphic or geographic ranges, but that most species might display some intermediate value.

It is the extreme cases—the radiations, the cases of bradytely, and the existence of trends—which traditionally have prompted detailed examination as macroevolutionary phenomena rather than "steady state." The reason seems to be that, in the steady state, the variables are quite "messy," and there is concomitant difficulty in making relatively precise, testable predictions about component species. But steady state seems to be by far the most common situation in macroevolution, and we include it here to show that it, too, can be understood (at least conceptually) in term of patterns of speciation and differential species survival. As an example, we cite the *Metacryphaeus* group of the calmoniid trilobites discussed above. The pattern is seen within the speciose genus *Metacryphaeus* itself (no real change accumulating within this genus through time) and within the entire group. The group, to be sure, is anatomically highly varied, but the limits of this variation are present from the earliest appearance of the group, suggesting a centripetal mode, with no further development of major anatomical specialization.

Trends

The final macroevolutionary topic listed in table 6.1 is trends. We have discussed trends at length (see pp. 266, 283) in our attempt to show that evolutionary phenomena can profitably be viewed as "decoupled" from microevolutionary processes. Consequently, we shall limit our discussion of this phenomenon here to a brief characterization of its properties, and to predictions which can be tested.

Eldredge and Gould (1972:111) and Schaeffer, Hecht, and Eldredge (1972:35) have noted that many trends seem to be selected a posteriori renderings of morphological sequences arranged in stratigraphic order. However, trends do exist: the comparative anatomy of anthropoid primates suggests that the present brain size of *Homo sapiens* is an enlargement over the plesiomorphic condition for hominids. Discounting a saltationist assumption, it might be predicted that, were an adequate sequence of hominid crania available, a correlation between progressive enlargement of cranial capacity and progressively younger sediments would be found. And this is what *is* found, at least in the overall pattern of the fossil record.

This and other similar trend patterns seem to involve low diversity, a speciation/extinction ratio of about 1, and moderate to high rates of speciation. The hypothesized mode of species selection is directional (the sense in which the expression "species selection" is most commonly used, e.g., Grant 1963, 1977; Stanley 1975; Gould 1977b; Gould and Eldredge 1977). Predictions include: (1) The species are stenotopic with respect to the structures and behaviors displaying the trend—i.e., these features are to be regarded as specializations pertinent to the actual occupation and exploitation of niches; (2) Stratigraphic ranges should be relatively short; (3) Geographic spread probably limited; and (4) Almost by definition, there should be a progression from relatively more plesiomorphic to relatively more apomorphic forms, coincident with the stratigraphic occurrence of the species. Actually, the latter "prediction" is the very criterion of recognition of a trend, and the pattern of species diversity can, in this case, perhaps be regarded as a prediction.

The *Probolops* group of calmoniids offers a test case. Earliest members develop a spine on the middle of the glabella (rather like a unicorn) and have thoracic and pygidial spines which emerge above the margin. Progressively up through the stratigraphic column, these spines occur nearer to the margin, so that in the youngest-known species (*Cryphaeoides rostratus*), the head spine is directly frontal, and the thoracic and pygidial spines are likewise entirely marginal. Whether the component species are stenotopic or not, their geographic and stratigraphic ranges are definitely short, and the structures involved are progressively apomorphous through time.

But perhaps the best test of the general species-selection hypothesis of trends is the lack of correlative changes within the stratigraphic ranges of component species. We did not list this as a prediction because (see pp. 285) an intraspecific trend may be coincident with an interspecific trend (i.e., in the same feature) and yet not, strictly speaking, falsify the hypothesis (or, of course, falsify the conventional syntheticist hypothesis either). But the converse holds: demonstration that an interspecific trend is not coincident with whatever pattern pertains within a species falsifies the hypothesis of continuity between microevolution and macroevolution. And, perhaps only by dint of lack of imagination, these are the only two sets of hypotheses we have been considering.

In any case, the *Probolops* group displays insufficient stratigraphic evidence to evaluate within-species patterns. But the hominid fossil record (figure 4.10) while agreeing with the general set of predictions (increase in brain size is here assumed to be adaptive and related to niche occupation and exploitation of hominid species), also demonstrates (a) considerable stratigraphic overlap between species with different cranial capacities and (b) long periods (i.e., the entire stratigraphic range of a given species) of little or no appreciable change in cranial capacity within a species. The trend is not, manifestly, a function of intraspecific, gradual, progressive, microevolutionary processes. We owe our braininess, it would seem, to directional species selection.

Systematics and the Evolutionary Process: A Summary

The main points we have argued in this chapter are as follows:

1. We have two basic disagreements with contemporary evolutionary theory. The first is methodological, and the second substantive. (i) Corroborated theories of relationship can be used as part of the data array to test predictions of evolutionary patterns stemming from theories of the evolutionary process. However, much of contemporary evolutionary theory, especially that pertaining to interspecific evolutionary phenomena, has not been formulated in a manner which renders evolutionary hypotheses easily susceptible to criticism. (ii) The standard syntheticist assumption is that processes considered important in changing gene content and frequency within populations, especially natural selection, account for all such changes in evolution. Within-population mechanisms are directly extrapolated to higher taxonomic levels in an effort to articulate such a connection. We do not argue here that within-population processes as envisioned in the neo-Darwinian paradigm are false. We recognize that the within-population concept of adaptation via natural selection is a viable hypothesis accounting for the deterministic part of much genetic, behavioral, and morphological change in evolution. However we strongly disagree that a smooth extrapolation of within-population (microevolutionary) processes is a logical or effective integration of such within-population mechanisms with among-species phenomena (macroevolution).

The major conceptual problem with such a direct "syntheticist" extrapolation of corroborated microevolutionary processes as mechanisms for the production of macroevolutionary patterns, is that it collides with a major, equally well-corroborated hypothesis of systematic biology: species are real, discrete entities in nature and have, among other things, origins, histories, and extinctions. The assumption that life has evolved, from the standpoint of systematics, implies that species evolve from other species. Theories of speciation abound. Speciation is not, fundamentally, a process of adaptation. Therefore, a theory emphasizing adaptation at its core cannot properly be extrapolated smoothly from the level of microevolution to the level of macroevolution. The conflict with the syntheticist form of macroevolutionary theory, which is a direct, wholesale extrapolation of within-species, microevolutionary theory arises from the necessary obliteration of species as the basic evolutionary units in the syntheticist view. In the transformational approach, the emphasis is on the (adaptive) change of the intrinsic features of a taxon into some other form, retrospectively classified as a new species, and perhaps a new genus, family, order, etc. This emphasis on the transformation of morphology as the basic question of evolution permeates not only syntheticist formulations of macroevolutionary theory, but also the writings of such nonconformists as Grassé, Goldschmidt, and others. Most such theories represent a direct fusion of population (and other areas of) genetics with the data and ideas of paleontology. We have dubbed all such formulations, syntheticist or not, the "transformational" approach to evolution.

The "taxic" approach, on the other hand, recognizes that species are evolutionary units. A macroevolutionary theory is incomplete if it does not take this generalization into account. We have reviewed the major contributions of both approaches to macroevolutionary theory, showing how the preponderance of work has argued for direct extrapolation of within-species phenomena as a realistic mechanism for the production of among-species phenomena, and also showing that some biologists have seen the conceptual difficulty if speciation theory is omitted.

2. We further argue that if evolutionary theory is to be improved and better integrated, its phenomenological components must first be recognized. We agree with some previous authors who recognize a distinction between the levels of (a) microevolution (change of gene content and frequency within species) and (b) macroevolution (change of species composition in time and space within a monophyletic group). Speciation is the process separating the two phenomenological levels.

With regard to these levels: (i) Hypotheses concerning microevolutionary phenomena can be tested experimentally and modelled

mathematically. The standard neo-Darwinian paradigm, in basic outline, seems well corroborated. Natural selection is the dynamic effecting the deterministic component of such change. Generational
data are needed to test such hypotheses; thus microevolutionary
theory and the phenomena associated with it (especially natural selection) are beyond the normal purview of systematic biology. (ii) Hypotheses concerning mechanisms of speciation require detailed
phylogenetic trees for testing—a problematical requirement. Nonetheless, the broad outlines of the mechanisms of speciation seem
fairly well worked out, mostly from pattern analysis by systematists
working in the field. (iii) Hypotheses concerning macroevolutionary
processes require a cladogram (i.e., a well-corroborated hypothesis
that a group is monophyletic) plus distributional data pertaining to
component species. The data of systematics seem most appropriate
to the testing of macroevolutionary hypotheses. The dynamic involved seems to be a process of differential species survival ("species selection"—an analogue, to be sure, of natural selection). In
terms of the Linnaean hierarchy, there is nothing more to macroevolution than species, inasmuch as taxa of higher rank than species do
not exist in the same sense as do species, and thus can in no way
be construed as evolutionary units; rather they are (when monophyletic, i.e., correctly delineated) expressions of the branching pattern
produced by many speciation events through time.

3. We then argue, following Stanley (1975), Salthe (in press),
and earlier workers, that, to some extent at least, the phenomenological levels of evolution are "decoupled," i.e., that each level
has its own internal dynamic and thus works at least partially independently of its neighboring upper or lower levels. Each level can
be understood on its own terms, without complete reference to the
terms of the next lower level. The levels are only partly reducible or
"decomposable." Arguments and test examples are presented in
support of the decoupling hypothesis.

However, once decoupling is established, we examine the nature of the interactions between the levels. The levels are by no
means totally divorced, and phenomena at one level are frequently
"resolved" at successively lower levels. Specifically, adaptation via
natural selection may account for all observed morphologies
throughout the biotic world since its inception; but it does so only in
the sense that species are always adapted to their own niche (by

definition, nearly), possessing anatomical, physiological, and behavioral properties appropriate to each individual niche. Interspecific interactions (e.g., competition) may be resolved, in part, by microevolutionary changes in sympatric populations of one or more of the species involved. At the higher level of macroevolution, each component species is assumed to be well adapted to its niche; the process of differential species survival therefore involves fundamentally a rearrangement of niche space through time. This may be worked out, in part, by microevolutionary change, but at base is patently *not* a matter of progressive adaptive change in gene content and frequency within a species in response to new or changing environmental conditions.

4. The basic outline of a testable theory of macroevolution is thus indicated. Such a theory requires (a) a well-corroborated cladogram (to insure that one is dealing with a monophyletic taxon and to evaluate distributions of plesiomorphies and apomorphies of component species) and (b) a pattern of distribution in space as well as in time (if available) of the individual species sampled. We recognize the strong "random" component of such shifts in diversity patterns within monophyletic groups through time, but argue that a deterministic view is a priori at least equally plausible.

We then present an outline of a specific macroevolutionary theory. It contains little that is new, its originality residing in its form. The major virtue we claim for the theory is that it is testable, i.e., predictions concerning component species of a monophyletic group exhibiting a particular pattern of diversity can be generated and tested, and that it explicitly incorporates the origin and extinction of species.

The theory draws on ecological niche theory as the conceptual nexus between morphology (adaptation) on the one hand, and species diversity (seen as an interplay between rates of speciation and rates of extinction) on the other hand. This is to say that both adaptation (as reflected in anatomical and behavioral specializations which species possess) and diversity (number of species living in a given area) are related to the occupation and exploitation of ecological niches. We then review some of the theoretical work on the controls of speciation and extinction. The central hypothesis, in this connection, is that broad-niched, relatively eurytopic species (a) tend to react to interspecific competition by mutual exclusion, (b) have low extinction rates (i.e., relatively long stratigraphic ranges), (c) tend to occur over wide geographic ranges (relative to the collective range of all of the species within the monophyletic group), and (d) retain plesiomorphies, i.e., species tend to resemble one another closely, and few if any species develop autapomorphies which can be hypothesized to be evolutionary specializations related to the occupation of a particular niche. Relatively stenotopic species (a) tend to react to interspecific competition by subdivision of niche space, (b) have high extinction rates, (c) exhibit relatively narrow geographic distributions, and (d) tend to develop anatomical specializations which appear as autapomorphies.

With the general notion of factors controlling rates of speciation and extinction in mind, we constructed a table (table 6.1) on which various macroevolutionary patterns ("adaptive radiations," "arrested evolution," "steady state," and "trends") are characterized in terms of (a) diversity patterns, (b) ratio of speciation to extinction, and (c) absolute rates of speciation as criteria for recognition of the pattern itself. We then give a conjectural characterization of the apparent mode of "species selection" for each macroevolutionary pattern. The theory is evaluated in light of the predictions concerning the component species which flow from it for each case. The predictions include geographic distributions, stratigraphic range (i.e., extinction rate of species), relative eurytopy or stenotopy of component species, and the distribution of plesiomorphous and apomorphous conditions within the species. We test the theory with particular cases relevant to the four macroevolutionary patterns tabulated. We do not claim that the theory is highly corroborated. Our main concern, rather, is to elaborate a theory of macroevolution consistent with the phenomenological levels we recognize in evolution—a theory which both explicitly incorporates the existence and evolution of species and which is in a form whereby it can be tested and either corroborated or rejected. Success in meeting these two objectives far outweighs the importance of the eventual fate of the theory itself.

References

Adanson, M. 1763. *Familles des plantes,* vol. 1. Vincent, Paris.

Agassiz, J. L. R. 1844. *Recherches sur les poissons fossiles.* Neuchâtel, Switzerland.

Anderson, D. T. 1973. *Embryology and Phylogeny in Annelids and Arthropods.* Pergamon Press, Oxford.

Ashlock, P. D. 1971. Monophyly and associated terms. *Syst. Zool.* 20:63–69.

Ashlock, P. D. 1972. Monophyly again. *Syst. Zool.* 21:430–37.

Ashlock, P. D. 1974. The uses of cladistics. *Ann. Rev. Ecol. Syst.* 5:81–99.

Ayala, F. J., ed. 1976. *Molecular Evolution.* Sinauer Ass., Sunderland, Mass.

Bassindale, R. 1936. The developmental stages of three English barnacles, *Balanus balanoides* (Linn.), *Chthamalus stellatus* (Poli), and *Verruca stroemia* (O.F. Müller). *Proc. Zool. Soc. London* 1936:57–74.

Bell, B. M. 1976. Phylogenetic implications of ontogenetic development in the Class Edrioasteroidea (Echinodermata). *J. Paleontol.* 50:1001–19.

Bock, W. J. 1963. The cranial evidence for ratite affinities. *Proc. 13th Int. Ornithol. Congr.,* vol. 1, pp. 39–54.

Bock, W. J. 1965. The role of adaptive mechanisms in the origin of higher levels of organization. *Syst. Zool.* 14:272–87.

Bock, W. J. 1970. Microevolutionary sequences as a fundamental concept in macroevolutionary models. *Evolution* 24:704–22.

Bock, W. J. 1972. Species interactions and macroevolution. *Evol. Biol.* 5:1–24.

Bock, W. J. 1973. Philosophical foundations of classical evolutionary classification. *Syst. Zool.* 22:375–92.

Bock, W. J. 1977. Foundations and methods of evolutionary classification. In M. K. Hecht, P. C. Goody, and B. M. Hecht (eds.), *Major Patterns in Vertebrate Evolution,* pp. 851–95. Plenum, New York.

Bock, W. J. 1979. The synthetic explanation of macroevolutionary change—a reductionistic approach. In J. H. Schwartz, and H. B. Rollins (eds.), *Models and Methodologies in Evolutionary Theory,* pp. 20–69. *Bull. Carnegie Mus. Nat. Hist.* 13.

Bock, W. J., and G. Von Wahlert. 1963. Two evolutionary theories—a discussion. *Brit. J. Phil. Sci.* 14:140–46.

Bonde, N. 1977. Cladistic classification as applied to vertebrates. In M. K.

332 References

Hecht, P. C. Goody, and B. M. Hecht (eds.), *Major Patterns in Vertebrate Evolution,* pp. 741–804. Plenum, New York.

Boucot, A. J. 1975. *Evolution and Extinction Rate Controls.* Elsevier, New York.

Boucot, A. J. 1978. Community evolution and rates of cladogenesis. *Evol. Biol.* 11:545–655.

Bretsky, P. W., and D. M. Lorenz. 1969. Adaptive response to environmental stability: a unifying concept in paleoecology. *Proc. North Amer. Paleontol. Conv.,* part E, pp. 522–50.

Bretsky, P. W., and D. M. Lorenz. 1970. An essay on genetic–adaptive strategies and mass extinctions. *Bull. Geol. Soc. Amer.* 81:2449–56.

Bretsky, S. S. 1979. Recognition of ancestor–descendant relationships in invertebrate paleontology. In J. Cracraft and N. Eldredge (eds.), *Phylogenetic Analysis and Paleontology,* pp. 113–63. Columbia Univ. Press, New York.

Brinkmann, R. 1929. Statistische-biostratigraphische Untersuchungen an mittel jurassischen Ammoniten über Artbegriff und Stammesentwicklung. *Abh. Ges. Wiss. Göttingen, Math-Phys. K1.,* N. F., vol. 8, pt. 3.

Brown, W. L., and E. O. Wilson. 1956. Character displacement. *Syst. Zool.* 5:49–64.

Brundin, L. 1966. Transantarctic relationships and their significance, as evidenced by chironomid midges. *Kungl. Svenska Vetenskap.* Handl. 11:1–472.

Buck, R. C., and D. L. Hull. 1966. The logical structure of the Linnaean hierarchy. *Syst. Zool.* 15:97–111.

Burkhardt, R. W., Jr. 1977. *The Spirit of System: Lamarck and Evolutionary Biology.* Harvard Univ. Press, Cambridge.

Bush, G. L. 1975. Modes of animal speciation. *Ann. Rev. Ecol. Syst.* 6:339–64.

Cain, A. J., 1954. *Animal Species and Their Evolution.* Hutchinson and Co., London. (Reprinted 1960, Harper, New York.)

Cain, A. J., and G. A. Harrison. 1960. Phyletic weighting. *Proc. Zool. Soc. London* 135:1–31.

Carruthers, R. G. 1910. On the evolution of *Zaphrentis delanouei* in Lower Carboniferous times. *Quart. J. Geol. Soc. London* 66:523–38.

Carson, H. L. 1975. The genetics of speciation at the diploid level. *Amer. Nat.* 109:83–92.

Coleman, W. 1964. *Georges Cuvier Zoologist.* Harvard Univ. Press, Cambridge, Mass.

Colless, D. H. 1977. A cornucopeia of categories. *Syst. Zool.* 26:349–52.

Cracraft, J. 1974a. Phylogeny and evolution of the ratite birds. *Ibis* 116:494–521.

Cracraft, J. 1974b. Phylogenetic models and classification. *Syst. Zool.* 23:71–90.

Cracraft, J. 1978. Comparative biology and brain evolution. *Syst. Zool.* 27:260–64.

Cracraft J. 1979. Phylogenetic analysis, evolutionary models and paleontology. In J. Cracraft and N. Eldredge (eds.), *Phylogenetic Analysis and Paleontology,* pp. 7–39. Columbia Univ. Press, New York.

Crowson, R. A. 1970. *Classification and Biology.* Heinemann Educational Books, London.

Crowson, R. A. 1972. A systematist looks at cytochrome c. *J. Mol. Evol.* 2:28–37.

Cuénot, L. 1932. *La Genèse des espèces animales.* Librairie Félix Alcan, Paris.

Darwin, C. 1859. *On the Origin of Species.* Facsimile ed., 1967, Atheneum, New York.

deBeer, G. 1958. *Embryos and Ancestors.* Oxford Univ. Press, Oxford.

Delson, E. 1977. Catarrhine phylogeny and classification: Principles, methods and comments. *J. Human Evol.* 6:433–59.

Delson, E., and P. Andrews. 1975. Evolution and interrelationships of the catarrhine primates. In W. P. Luckett, and F. S. Szalay (eds.), *Phylogeny of the Primates,* pp. 405–46. Plenum, New York.

Delson, E., N. Eldredge, and I. Tattersall. 1977. Reconstruction of hominid phylogeny: A testable framework based on cladistic analysis. *J. Human Evol.* 6:263–78.

Dobzhansky, T. 1937. *Genetics and the Origin of Species.* Columbia Univ. Press, New York.

Dobzhansky, T. 1941. *Genetics and the Origin of Species,* 2nd rev. ed., Columbia Univ. Press, New York.

Dobzhansky, T. 1951. *Genetics and the Origin of Species,* 3rd rev. ed., Columbia Univ. Press, New York.

Dobzhansky, T. 1970. *Genetics of the Evolutionary Process.* Columbia Univ. Press, New York.

Dobzhansky, T., F. J. Ayala, G. L. Stebbins, and J. W. Valentine. 1977. *Evolution.* W. H. Freeman, San Francisco.

Ehrlich, P. R., and A. H. Ehrlich. 1967. The phenetic relationships of the butterflies: I. Adult taxonomy and the nonspecificity hypothesis. *Syst. Zool.* 16:301–17.

Ehrlich, P. R., and P. H. Raven. 1969. Differentiation of populations. *Science* 165:1228–32.

Eldredge, N. 1968. Convergence between two Pennsylvanian gastropod species: A multivariate mathematical approach. *J. Paleontol.* 42:186–96.

Eldredge, N. 1971. The allopatric model and phylogeny in Paleozoic invertebrates. *Evolution* 25:156–67.

Eldredge, N. 1972a. Systematics and evolution of *Phacops rana* (Green 1832) and *Phacops iowensis* Delo, 1935 (Trilobita) from the Middle Devonian of North America. *Bull. Amer. Mus. Nat. Hist.* 147:45–114.

Eldredge, N. 1972b. Morphology and relationships of *Bouleia* Kozlowski, 1923 (Trilobita, Calmoniidae). *J. Paleontol.* 46:140–151.

Eldredge, N. 1974a. Stability, diversity and speciation in Paleozoic epeiric seas. *J. Paleontol.* 48:540–548.

Eldredge, N. 1974b. Testing evolutionary hypotheses in paleontology: A comment on Makurath and Anderson (1973). *Evolution* 28:479–81.

Eldredge, N. 1975. Survivors from the good old, old, old days. *Nat. Hist.* 84(2):60–69.

Eldredge, N. 1979a. Cladism and common sense. In J. Cracraft and N. Eldredge (eds.), *Phylogenetic Analysis and Paleontology,* pp. 165–97. Columbia Univ. Press, New York.

Eldredge, N. 1979b. Alternative approaches to evolutionary theory. In J. H. Schwartz and H. B. Rollins (eds.), *Models and Methodologies in Evolutionary Theory,* pp. 7–19; *Bull. Carnegie Mus. Nat. Hist.* 13.

Eldredge, N., and L. Braniša. 1980. Calmoniid trilobites of the Lower Devonian *Scaphiocoelia* Zone of Bolivia, with remarks on related species. *Bull. Amer. Mus. Nat. Hist.* 165(2):181–290.

Eldredge, N. and J. Cracraft. 1979. Introduction to the symposium. In J. Cracraft and N. Eldredge (eds.), *Phylogenetic Analysis and Paleontology,* pp. 1–5. Columbia Univ. Presss, New York.

Eldredge, N., and M. J. Eldredge. 1972. A trilobite odyssey. *Nat. Hist.* 81(10):52–59.

Eldredge, N., and S. J. Gould. 1972. Punctuated equilibria: An alternative to phyletic gradualism. In T. J. M. Schopf (ed.), *Models in Paleobiology,* pp. 82–115. Freeman, Cooper and Co., San Francisco.

Eldredge, N., and S. J. Gould. 1974. Reply to Hecht. *Evol. Biol.* 7:303–8.

Eldredge, N., and A. R. Ormiston. 1979. Biogeography of Silurian and Devonian trilobites of the Malvinokaffric Realm. In J. Gray, and A. J. Boucot (eds.), *Historical Biogeography, Plate Tectonics and the Changing Environment,* pp. 147–67 Oreg. State Univ. Press, Corvallis.

Eldredge, N., and I. Tattersall. 1975. Evolutionary models, phylogenetic reconstruction, and another look at hominid phylogeny. In F. S. Szalay (ed.), *Approaches to Primate Paleobiology; Contr. Primatol.* 5:218–42.

Englemann, G. F., and E. O. Wiley. 1977. The place of ancestor–descendant relationships in phylogeny reconstruction. *Syst. Zool.* 26:1–11.

Farris, J. S. 1974. Formal definitions of paraphyly and polyphyly. *Syst. Zool.* 23:548–54.

Farris, J. S. 1976. Phylogenetic classification of fossils with recent species. *Syst. Zool.* 25:271–82.

Farris, J. S. 1977. On the phenetic approach to vertebrate classification. In M. K. Hecht, P. C. Goody, and B. M. Hecht (eds.), *Major Patterns in Vertebrate Evolution,* pp. 823–50. Plenum, New York.

Fryer, G., and T. D. Iles. 1969. Alternative routes to evolutionary success as exhibited by African cichlid fishes of the genus *Tilapia* and the species flocks of the great lakes. *Evolution* 23:359–69.

Gaffney, E. S. 1975. A phylogeny and classification of the higher categories of turtles. *Bull. Amer. Mus. Nat. Hist.* 155:387–436.

Gaffney, E. S. 1977. The side-necked turtle family Chelidae: A theory of relationships using shared derived characters. *Amer. Mus. Novitates* No. 2620:1–28.

Gaffney, E. S. 1979. An introduction to the logic of phylogeny reconstruction. In J. Cracraft and N. Eldredge (eds.), *Phylogenetic Analysis and Paleontology,* pp. 79–111. Columbia Univ. Press, New York.

Ghiselin, M. T. 1974. A radical solution to the species problem. *Syst. Zool.* 23:536–44.

Ghiselin, M. T. 1977. On paradigms and the hypermodern species concept. *Syst. Zool.* 26:437–38.

Ghiselin, M. T., and L. Jaffe. 1973. Phylogenetic classification in Darwin's "Monograph on the Sub-Class Cirrepedia." *Syst. Zool.* 22:132–40.

Gingerich, P. D. 1976. Paleontology and phylogeny: patterns of evolution at the species level in Early Tertiary mammals. *Amer. J. Sci.* 276:1–28.

Gingerich, P. D. 1979. The stratophenetic approach to phylogeny reconstruction in vertebrate paleontology. In J. Cracraft and N. Eldredge (eds.), *Phylogenetic Analysis and Paleontology,* pp. 41–77. Columbia Univ. Press, New York.

Gingerich, P. D., and M. Schoeninger. 1977. The fossil record and primate phylogeny. *J. Human Evol.* 6:484–505.

Goldschmidt, R. 1940. *The Material Basis of Evolution.* Yale Univ. Press, New Haven, Conn.

Gorman, G. C. 1968. The relationships of *Anolis* of the *roquet* species group (Sauria: Iguanidae): III., Comparative study of display behavior. *Breviora,* no. 284, pp. 1–31.

Gould, S. J. 1977a. *Ontogeny and Phylogeny.* Belknap Press of Harvard Univ. Press, Cambridge, Mass.

Gould, S. J. 1977b. Eternal metaphors of paleontology. In A. Hallam (ed.), *Patterns of Evolution, as Illustrated by the Fossil Record,* pp. 1–26. Elsevier, New York.

Gould, S. J., and N. Eldredge. 1977. Punctuated equilibria: the tempo and mode of evolution reconsidered. *Paleobiology* 3:115–51.

Gould, S. J., D. M. Raup, J. J. Sepkoski Jr., T. J. M. Schopf, and D. S. Simberloff. 1977. The shape of evolution: a comparison of real and random clades. *Paleobiology* 3:23–40.

Grant, P. R. 1972. Convergent and divergent character displacement. *Biol. J. Linn. Soc.* 4:39–68.

Grant, V. 1963. *The Origin of Adaptations.* Columbia Univ. Press, New York.

Grant, V. 1977. *Organismic Evolution.* W. H. Freeman, San Francisco.

Grassé, P. P. 1973. *L'Évolution du vivant.* Albin Michel, Paris.

Greenbaum, I. F., and R. J. Baker. 1976. Evolutionary relationships in *Macrotus* (Mammalia; Chiroptera); biochemical variation and karyology. *Syst. Zool.* 25:15–25.

336 References

Greenwood, P. H. 1974. The cichlid fishes of Lake Victoria, East Africa: the biology and evolution of a species flock. *Bull. Brit. Mus. Nat. Hist.,* suppl. 6, pp. 1–134.

Greenwood, P H., R. S. Miles, and C. Patterson, eds. 1973. *Interrelationships of Fishes.* Academic Press, London.

Grene, M. 1959. Two evolutionary theories. *Brit. J. Phil. Sci.* 9:110–27, 185–94.

Griffiths, G. C. D. 1973. Some fundamental problems in biological classification. *Syst. Zool.* 22:338–43.

Haldane, J. B. S. 1932. *The Causes of Evolution.* Harper, New York.

Hampé, A. 1959. Contribution a l'étude du développement et de la régulation des déficiences et des excédents dans la patte de l'embryon de poulet. *Arch. Anat. Microsc. Morph. Exp.* 48:345–78.

Harper, C. W., Jr. 1976. Phylogenetic inference in paleontology. *J. Paleontol.* 50:180–93.

Hecht, M. K. 1965. The role of natural selection and evolutionary rates in the origin of higher levels of organization. *Syst. Zool.* 14:301–17.

Hecht, M. K. 1976. Phylogenetic inference and methodology as applied to the vertebrate record. *Evol. Biol.* 9:335–63.

Hecht, M. K., and J. L. Edwards. 1976. The determination of parallel or monophyletic relationships: the proteid salamanders—a test case. *Amer. Nat.* 110:653–77.

Hecht, M. K., and J. L. Edwards. 1977. The methodology of phylogenetic inference above the species level. In M. K. Hecht, P. C. Goody, and B. M. Hecht (eds.), *Major Patterns of Vertebrate Evolution,* pp. 3–51. Plenum, New York.

Hennig, W. 1950. *Grundzüge einer Theorie der phylogenetischen Systematik.* Deutscher Zentralverlag, Berlin.

Hennig, W. 1965. Phylogenetic systematics. *Ann. Rev. Ent.* 10:97–116.

Hennig, W. 1966. *Phylogenetic Systematics.* Univ. Ill. Press, Urbana.

Hennig, W. 1969. *Die Stammesgeschichte der Insekten.* Waldemar Kramer, Frankfurt am Main.

Hennig, W. 1975. Cladistic analysis or cladistic classification?—a reply to Ernst Mayr. *Syst. Zool.* 24:244–56.

Hopwood, A. T. 1950a. Animal classification from the Greeks to Linnaeus. In *Lectures on the Development of Taxonomy,* pp. 24–32. Linnean Soc., London.

Hopwood, A. T. 1950b. Animal classification from Linnaeus to Darwin. In *Lectures on the Development of Taxonomy,* pp. 46–59. Linnean Soc., London.

Hull, D. L. 1964. Consistency and monophyly. *Syst. Zool.* 13:1–11.

Hull, D. L. 1965. The effect of essentialism on taxonomy—two thousand years of stasis (1). *Brit. J. Phil. Sci.* 15:2–14.

Hull, D. L. 1976. Are species really individuals? *Syst. Zool.* 25:174–91.

Hull, D. L. 1978. A matter of individuality. *Phil. Sci.* 45:335–60.

Huxley, J. S. 1958. Evolutionary processes and taxonomy with special reference to grades. *Uppsala Univ. Arssk.* 1958:21–38.

Johnson, L. A. S. 1970. Rainbow's end: the quest for an optimal taxonomy. *Syst. Zool.* 19:203–39.

Kaestner, A. 1970. *Invertebrate Zoology,* vol. 3: *Crustacea.* John Wiley and Sons, New York.

Kahl, M. P. 1971. Social behavior and taxonomic relationships of the storks. *Living Bird* 10:151–70.

Kahl, M. P. 1972. Comparative ethology of the Ciconiidae. The wood-storks (genera *Mycteria* and *Ibis*). *Ibis* 114:15–29.

Kent, G. C., Jr. 1965. *Comparative anatomy of the vertebrates.* C. V. Mosby Co., St. Louis, Mo.

Kerkut, G. A. 1960. *The Implications of Evolution.* Pergamon, New York.

Kollar, E. J. 1972. The development of the integument: spatial, temporal and phylogenetic factors. *Amer. Zool.* 12:125–35.

Lack, D. 1947. *Darwin's finches.* Cambridge Univ. Press, Cambridge. (Reprinted 1961 as Harper Torchbook.)

Lamarck, J. B. 1914. *Zoological Philosophy.* H. Elliot, Trans. MacMillan and Co., London.

Larson, J. L. 1968. Linnaeus and the natural method. *Isis* 58:304–20.

Lewontin, R. C. 1978. Adaptation. *Sci. Amer.* 239:212–30.

Linnaeus, Carolus (Carl von Linné). 1758. *Caroli Linnaei Systema Naturae, Regnum Animale,* 10th ed. Brit. Mus. (Nat. Hist.), London. (Reprinted 1939.)

Løvtrup, S. 1977. *The Phylogeny of Vertebrata.* John Wiley and Sons, New York.

MacArthur, R. H., and E. O. Wilson. 1967. *The Theory of Island Biogeography.* Princeton Univ. Press, Princeton, N.J.

MacFadden, B. J. 1976. Cladistic analysis of primitive equids, with notes on other perissodactyls. *Syst. Zool.* 25:1–14.

McKenna, M. C. 1975. Toward a phylogenetic classification of the Mammalia. In W. P. Lucket and F. S. Szalay (eds.), *Phylogeny of the Primates,* pp. 21–46. Plenum, New York.

Makurath, J. H., and E. J. Anderson. 1973. Intra- and interspecific variation in gypidulid brachiopods. *Evolution* 27:303–10.

Margulis, L. 1974. Five-kingdom classification and the origin and evolution of cells. *Evol. Biol.* 7:45–78.

Maslin, T. P. 1952. Morphological criteria of phyletic relationships. *Syst. Zool.* 1:49–70.

Mayr, E. 1940. Speciation phenomena in birds. *Amer. Nat.* 74:249–78.

Mayr, E. 1942. *Systematics and the Origin of Species.* Columbia Univ. Press, New York. (Dover edition, 1964.)

Mayr, E. 1963. *Animal Species and Evolution.* Harvard Univ. Press, Cambridge, Mass.

Mayr, E. 1969. *Principles of Systematic Zoology.* McGraw-Hill, New York.

338 References

Mayr, E. 1970. *Populations, Species and Evolution.* Harvard Univ. Press, Cambridge, Mass.

Mayr, E. 1974. Cladistic analysis or cladistic classification? *Z. Zool. Syst. Evolut.-forsch.* 12:94–128.

Meise, W. 1963. Verhalten der Straussartigen Vögel und Monophylie der Ratitae. *Proc. 13th Ornith. Congr.* 1:115–25.

Michener, C. D. 1978. Dr. Nelson on taxonomic methods. *Syst. Zool.* 27:112–18.

Minkoff, E. C. 1965. The effects on classification of slight alterations in numerical technique. *Syst. Zool.* 14:196–213.

Morton, J. E. 1958. *Molluscs.* Hutchinson and Co., London (Reprinted, 1960, Harper, New York).

Moy-Thomas, J. A., and R. S. Miles. 1971. *Paleozoic Fishes.* W. B. Saunders, Philadelphia.

Nelson, G. J. 1969. Gill arches and the phylogeny of fishes, with notes on the classification of vertebrates. *Bull. Amer. Mus. Nat. Hist.* 141:475–552.

Nelson, G. L. 1970. Outline of a theory of comparative biology. *Syst. Zool.* 19:373–84.

Nelson, G. L. 1971a. "Cladism" as a philosophy of classification. *Syst. Zool.* 20:373–76.

Nelson, G. J. 1971b. Paraphyly and polyphyly: redefinitions. *Syst. Zool.* 20:471–72.

Nelson, G. J. 1972a. Comments on Hennig's "Phylogenetic Systematics" and its influence on ichthyology. *Syst. Zool.* 21:364–74.

Nelson, G. J. 1972b. Phylogenetic relationship and classification. *Syst. Zool.* 21:227–31.

Nelson, G. J. 1973a. The higher-level phylogeny of vertebrates. *Syst. Zool.* 22:87–91.

Nelson, G. J. 1973b. Monophyly again?—a reply to P. D. Ashlock. *Syst. Zool.* 22:310–12.

Nelson, G. J. 1973c. Classification as an expression of phylogenetic relationships. *Syst. Zool.* 22:344–59.

Nelson, G. J. 1974. Darwin–Hennig classification: a reply to Ernst Mayr. *Syst. Zool.* 23:452–58.

Nelson, G. J. 1978. Ontogeny, phylogeny, paleontology, and the biogenetic law. *Syst. Zool.* 27:324–45.

Nelson, G. J. 1979. Cladistic analysis and synthesis: principles and definitions, with a historical note on Adanson's "Familles des plantes" (1763–1764). *Syst. Zool.* 28:1–21.

Nelson, G. J., and N. I. Platnick. 1980. *Cladistics and Vicariance: Patterns in Comparative Biology.* Columbia Univ. Press, New York.

Nelson, G. J., and D. E. Rosen, eds. 1980. *Vicariance biogeography: Congruence of earth history with plant and animal distributions.* Columbia Univ. Press, New York.

Parkes, K. C., and G. A. Clark, Jr. 1966. An additional character linking ratites and tinamous, and an interpretation of their monophyly. *Condor* 68:459–71.

Patterson, C. 1977. The contribution of paleontology to teleostean phylogeny. In M. K. Hecht, P. C. Goody, and B. M. Hecht (eds.), *Major Patterns in Vertebrate Evolution,* pp. 579–643. Plenum, New York.

Patterson, C., and D. E. Rosen. 1977. Review of icthyodectiform and other Mesozoic teleost fishes and the theory and practice of classifying fossils. *Bull. Amer. Mus. Nat. Hist.* 158:81–172.

Platnick, N. I. 1977a. Paraphyletic and polyphyletic groups. *Syst. Zool.* 26:195–200.

Platnick, N. I. 1977b. Cladograms, phylogenetic trees, and hypothesis testing. *Syst. Zool.* 26:438–42.

Platnick, N. I. 1977c. The hypochiloid spiders: a cladistic analysis, with notes on the Atypoidea (Arachnida, Araneae). *Amer. Mus. Novitates* 2627:1–23.

Platnick, N. I. 1978a. Classifications, historical narratives and hypotheses. *Syst. Zool.* 27:365–69.

Platnick, N. I. 1978b. Gaps and prediction in classification. *Syst. Zool.* 27:472–74.

Platnick, N. I., and W. J. Gertsch. 1976. The suborders of spiders: a cladistic analysis (Arachnida, Araneae). *Amer. Mus. Novitates* 2607:1–15.

Platnick, N. I., and M. U. Shadab. 1976. A revision of the neotropical spider genus *Zimiromus,* with notes on *Echemus* (Araneae, Gnaphosidae). *Amer. Mus. Novitates* 2609:1–24.

Popper, K. R. 1959. *The Logic of Scientific Discovery.* Harper Torchbooks, New York.

Prosser, C. L., and F. A. Brown, Jr. 1965. *Comparative Animal Physiology.* W. B. Saunders, Philadelphia.

Raikow, R. J. 1975. The evolutionary reappearance of ancestral muscles as developmental anomalies in two species of birds. *Condor* 77:514–17.

Raup, D. M. 1966. Geometric analysis of shell coiling: general problems. *J. Paleontol.* 40:1178–90.

Raup, D. M. 1972. Taxonomic diversity during the Phanerozoic. *Science* 177:1065–71.

Raup, D. M. 1977. Stochastic models in evolutionary paleontology. In A. Hallam (ed.) *Patterns of Evolution, as illustrated by the Fossil Record,* pp. 59–78. Elsevier, New York.

Raup, D. M., and S. J. Gould. 1974. Stochastic simulation and evolution of morphology—towards a nomothetic paleontology. *Syst. Zool.* 23:305–22.

Raup, D. M., S. J. Gould, T. J. M. Schopf, and D. S. Simberloff. 1973. Stochastic models of phylogeny and the evolution of diversity. *J. Geol.* 81:525–42.

Rensch, B. 1960. *Evolution Above the Species Level.* Columbia Univ. Press, New York.

Romanes, G. J. 1886. Physiological selection: an additional suggestion on the origin of species. *J. Linn. Soc. Zool.* 19:337–411.

340 References

Romer, A. S. 1962. *The Vertebrate Body.* W. B. Saunders, Philadelphia.

Romer, A. S. 1966. *Vertebrate Paleontology.,* 3rd ed. Univ. of Chicago Press, Chicago.

Rosa, D. 1931. *L'Ologenèse; nouvelle théories de l'évolution et de la distribution geographique des êtres vivants.* Librairie Félix Alcan, Paris.

Ross, H. H. 1974. *Biological Systematics.* Addison-Wesley, Reading, Mass.

Rowe, A. W. 1899. An analysis of the genus *Micraster* as determined by rigid zonal collecting from the zone of *Rhynchonella Cuvieri* to that of *Micraster cor-anguinum. Quart. J. Geol. Soc. London* 55:494–547.

Salthe, S. N. 1975. Problems of macroevolution (molecular evolution, phenotype definition and canalization) as seen from an hierarchical viewpoint. *Amer. Zool.* 15:295–314.

Salthe, S. N. In press. An integrated view of evolutionary theory.

Schaeffer, B., M. K. Hecht, and N. Eldredge. 1972. Phylogeny and paleontology. *Evol. Biol.* 6:31–46.

Schindewolf, O. H. 1950. *Grundfragen de Paläontologie.* Stuttgart, Schweizerbart.

Schnell, G. D. 1970. A phenetic study of the suborder Lari (Aves): 1. Methods and results of principal components analyses. *Syst. Zool.* 19:35–57.

Shaw, A. B. 1964. *Time in Stratigraphy.* McGraw-Hill, New York.

Shear, W. A. 1975. The opilinoid family Caddidae in North America, with notes on species from other regions (Opiliones, Palpatores, Caddoidea). *J. Arachnol.* 2:65–88.

Simpson, G. G. 1944. *Tempo and Mode in Evolution.* Columbia Univ. Press, New York.

Simpson, G. G. 1945. The principles of classification and a classification of the Mammalia. *Bull. Amer. Mus. Nat. Hist.* 85:1–350.

Simpson, G. G. 1951. The species concept. *Evolution* 5:285–98.

Simpson, G. G. 1953. *The Major Features of Evolution.* Columbia Univ. Press, New York.

Simpson, G. G. 1959a. Mesozoic mammals and the polyphyletic origin of mammals. *Evolution* 13:405–14.

Simpson, G. G. 1959b. The nature and origin of supraspecific taxa. *Cold Spring Harbor Symp. Quant. Biol.* 24:255–71.

Simpson, G. G. 1959c. Anatomy and morphology: classification and evolution: 1859 and 1959. *Proc. Amer. Phil. Soc.* 103:286–306.

Simpson, G. G. 1961. *Principles of Animal Taxonomy.* Columbia Univ. Press, New York.

Simpson, G. G. 1963. The meaning of taxonomic statements. In S. L. Washburn (ed.), *Classification and Human Evolution,* pp. 1–31. Aldine, Chicago.

Simpson, G. G. 1975. Recent advances in methods of phylogenetic inference. In W. P. Luckett, and F. S. Szalay (eds.), *Phylogeny of the Primates, A Multidisciplinary Approach,* pp. 3–19. Plenum, New York.

Simpson, G. G. 1976. The compleat paleontologist? *Ann. Rev. Earth Planet. Sci.* 4:1–13.

Simpson, G. G. 1978. Review of "Patterns of Evolution, as Illustrated by the Fossil Record," edited by A. Hallam. *Nature* 273:77–78.

Sneath, P. H. A., and R. R. Sokal. 1973. *Numerical Taxonomy*. W. H. Freeman, San Francisco.

Sokal, R. R., and J. H. Camin. 1965. The two taxonomies: areas of agreement and conflict. *Syst. Zool.* 14:176–95.

Sokal, R. R., and T. J. Crovello. 1970. The biological species concept: a critical evaluation. *Amer. Nat.* 104:127–53.

Sokal, R. R., and F. J. Rohlf. 1962. The comparison of dendrograms by objective methods. *Taxon* 11:33–40.

Sokal, R. R., and P. H. A. Sneath. 1963. *Principles of Numerical Taxonomy*. W. H. Freeman, San Franscisco.

Stanley, S. M. 1975. A theory of evolution above the species level. *Proc. Nat. Acad. Sci.* 72:646–50.

Stenzel, H. B. 1949. Successional speciation in paleontology: the case of the oysters of the *sellaeformis* stock. *Evolution* 3:34–50.

Storer, T. I., and R. L. Usinger. 1965. *General Zoology*. McGraw-Hill, New York.

Szalay, F. S. 1977. Ancestors, descendants, sister groups and testing of phylogenetic hypotheses. *Syst. Zool.* 26:12–18.

Tattersall, I., and N. Eldredge. 1977. Fact, theory and fantasy in human paleontology. *Amer. Sci.* 65:204–11.

Trueman, A. E. 1922. The use of *Gryphaea* in the correlation of the Lower Lias. *Geol. Mag.* 59:256–68.

Valentine, J. W. 1969. Patterns of taxonomic and ecological structure of the shelf benthos during Phanerozoic time. *Palaeontology* 12:684–709.

Valentine, J. W. 1973. *Evolutionary Paleoecology of the Marine Biosphere*. Prentice-Hall, Englewood Cliffs, N.J.

Van Valen, L. 1973. A new evolutionary law. *Evol. Theory* 1:1–30.

Van Valen, L. 1978. Review of "Patterns of Evolution, as Illustrated by the Fossil Record," edited by A. Hallam. *Paleobiology* 4:210–17.

Waagen, W. 1869. Die Formenreihe des Ammonites subradiatus. *Geogn.-Paleont. Beit.* 2:181–256.

Waddington, C. H. 1967. Comment made during discussion of paper by Dr. Eden. In P. S. Moorehead and M. M. Kaplan (eds.), *Mathematical Challenges to the Neo-Darwinian Interpretation of Evolution*, p. 14. Wistar Institute Press, Philadelphia.

Wagner, M. 1869. *Die Enstehung der Arten durch räumliche Sonderung*. Benno Schwalbe, Basel. (Not seen.)

White, M. J. D. 1968. Models of speciation. *Science* 159:1065–70.

Wiley, E. O. 1975. Karl R. Popper, systematics, and classification: A reply to Walter Bock and other evolutionary taxonomists. *Syst. Zool.* 24:233–43.

Wiley, E. O. 1976. *The phylogeny and biogeography of fossil and Recent gars (Actinopterygii: Lepisosteidae)*. Univ. Kans. Mus. Nat. Hist. Misc. Publ. 64:1–111.

Wiley, E. O. 1978. The evolutionary species concept reconsidered. *Syst. Zool.* 27:17–26.

Wiley, E. O. 1979. Ancestors, species, and cladograms—remarks on the symposium. In J. Cracraft and N. Eldredge (eds.), *Phylogenetic Analysis and Paleontology,* pp. 211–25. Columbia Univ. Press, New York.

Williams, G. C. 1966. *Adaptation and Natural Selection.* Princeton Univ. Press, Princeton, N.J.

Williams, H. S. 1910. The migration and shifting of Devonian faunas. *Popular Sci. Monthly* 77:70–77.

Willis, J. C. 1940. *The Course of Evolution.* Cambridge Univ. Press. Cambridge.

Wilmot, A. J. 1950. Systematic botany from Linnaeus to Darwin. In *Lectures on the Development of Taxonomy,* pp. 33–45. Linnaean Soc., London.

Winsor, M. P. 1976. *Starfish, Jellyfish, and the Order of Life.* Yale Univ. Press, New Haven, Conn.

Wolf, A. 1930. *Textbook of Logic.* Collier Books, New York.

Wright, S. 1931. Evolution in mendelian populations. *Genetics* 16:97–159.

Wright, S. 1932. The roles of mutation, inbreeding, crossbreeding, and selection in evolution. *Proc. VI Int. Congr. Genetics,* 1:356–66.

Wright, S. 1945. Tempo and mode in evolution: a critical review. (Review of "Tempo and Mode in Evolution" by G. G. Simpson.) *Ecology* 26:415–19.

Wright, S. 1956. Modes of selection. *Amer. Nat.* 90:5–24.

Wright, S. 1967. Comments on the preliminary working papers of Eden and Waddington. In P. S. Moorehead and M. M. Kaplan (eds.) *Mathematical Challenges to the Neo-Darwinian Interpretation of Evolution,* pp. 117–120. Wistar Institute Press, Philadelphia.

Index